ADVANCES IN APPLIED SCIENCE AND ENGINEERING

# INNOVATIONS AND CHALLENGES IN MODERN PHYSICAL CHEMISTRY

## RESEARCH AND PRACTICES

# ADVANCES IN APPLIED SCIENCE AND ENGINEERING

Additional books and e-books in this series can be found on Nova's website under the Series tab.

ADVANCES IN APPLIED SCIENCE AND ENGINEERING

# INNOVATIONS AND CHALLENGES IN MODERN PHYSICAL CHEMISTRY

## RESEARCH AND PRACTICES

A. K. HAGHI
EDITOR

nova
science publishers
New York

Additional color graphics may be available in the e-book version of this book.

## Library of Congress Cataloging-in-Publication Data

Names: Haghi, A. K., editor.
Title: Innovations and challenges in modern physical chemistry: research and practices / A.K. Haghi (editor), Professor Emeritus of Engineering Sciences, University of Guilan, Iran.
Identifiers: LCCN 2020027181 (print) | LCCN 2020027182 (ebook) | ISBN 9781536181814 (hardcover) | ISBN 9781536182354 (adobe pdf)
Subjects: LCSH: Chemistry, Physical and theoretical.
Classification: LCC QD453.3 .I56 2020 (print) | LCC QD453.3 (ebook) | DDC 541--dc23
LC record available at https://lccn.loc.gov/2020027181
LC ebook record available at https://lccn.loc.gov/2020027182

*Published by Nova Science Publishers, Inc. † New York*

# CONTENTS

# PREFACE

This new volume, introduces readers to some of the latest research applications of physical chemistry. The compilation of this volume was motived by the tremendous increase of useful research work in the field of physical chemistry and related subjects in recent years, and the need for communication between physical chemists, physicists, and biophysicists. This volume reflects the huge breadth and diversity in research and the applications in physical chemistry and physical chemistry techniques, providing case studies that are tailored to particular research interests. It examines the industrial processes for emerging materials, determines practical use under a wide range of conditions, and establishes what is needed to produce a new generation of materials.

*A. K. Haghi, PhD*
Professor Emeritus of Engineering Sciences
University of Guilan, Iran

In: Innovations and Challenges in Modern … ISBN: 978-1-53618-181-4
Editor: A. K. Haghi 

*Chapter 1*

# PAIN FACE, SCIENCE AND ETHICS, HOT, SPICY, FRESH, ASPIRIN, OPIOIDS AND LAWS

***Francisco Torrens[1,*] and Gloria Castellano[2]***
[1]Institut Universitari de Ciència Molecular, Universitat de València, Edifici d'Instituts de Paterna, València, Spain
[2]Departamento de Ciencias Experimentales y Matemáticas, Facultad de Veterinaria y Ciencias Experimentales, Universidad Católica de Valencia San Vicente Mártir, València, Spain

## ABSTRACT

*Pain* is an unpleasant sensory and emotional experience associated with actual or potential tissue damage, or described in terms of such damage (International Association for the Study of Pain). Four types of pain exist: cognitive, physiological (opiate related), social and emotional ones. Three types of receptors exist: μ-, δ- and κ-ones. Therapeutics: opioid rotation. Watching: Analgesic, Activity, Aberrant behaviour and Adverse effects (the four As). Pain is avoidable: Chronic pain has no age in the 21$^{st}$ century. The predominant pain in the elderly is chronic pain. Change of paradigm in pain treatment. Mesenchymal stromal cells are advanced

---

* Corresponding Author's Email: torrens@uv.es.

drugs. Drug-checking services and European Union's early-warning system play an important role in reducing the harm associated with adulteration in the drug illegal market. It is basic to be at street level preventing. One should adapt himself to leisure and black market. Gender is important. Some drugs are legal and others, not *for economic-political and cultural reasons*.

**Keywords:** new psychoactive substance, adulteration, controlled drug, illegal market, pain pharmacological treatment, opioid analgesic, pain avoidability

## 1. Introduction

Setting the scene: the face of pain, science, ethics of a universal symptom, hot, spicy, fresh, salicylic acid (SA), acetylsalicylic acid (ASA), *aspirin*, *N*-acetyl-*p*-aminophenol (*paracetamol*), diazepam, morphine, diacetylmorphine, stimulants, adrenaline, amphetamine, caffeine (Caff), tetrahydrocannabinol (THC), dopamine, 3,4-dihydroxytyrosine, cocaine, nicotine, serotonin (5-HT), fluoxetin, methylenedioxymethamphetamine (MDMA), advances in pain pharmacological treatment with opioid analgesics, chronic pain having no age in the 21$^{st}$ century, drugs in Society 4.0, this being what it is, laws and drugs.

The International Association for the Study of Pain (IASP) defined pain: *Pain* is an unpleasant sensory and emotional experience associated with actual or potential tissue damage, or described in terms of such damage. There are four types of pain: cognitive, physiological, social and emotional pains: In particular, physiological pain results opiate related. There are three types of pain receptors: μ-, δ- and κ-receptors. Some treatments are based on opioid rotation. The four As of watching: Analgesic, Activity, Aberrant behaviour and Adverse effects (AEs). Pain results avoidable: Chronic pain has no age in the 21$^{st}$ century. The predominant pain in the elderly is chronic pain. Change of paradigm in pain treatment. Multipotent mesenchymal stromal cells (MMSCs) are advanced therapy medicines. Drug-checking services and EU's early-warning system can play an important role in reducing the harm associated with adulteration in the drug illegal market. It is essential to be at street level doing prevention. One should adapt himself to leisure and black market. The importance of gender. Some drugs are legal and some others are not *for economic-political and cultural reasons*. Journal *Mètode* published a special issue on pain entitled *The face of pain: Science and ethics of a universal symptom* (*cf.* Figure 1).

Figure 1. *Mètode* 2011 (71) special issue: The face of pain: Science and ethics of a universal symptom.

Earlier publications in *Nereis*, etc. classified yams [1], lactic acid bacteria [2], fruits [3], spices [4] and oil legumes [5] by principal component, cluster and meta-analysis. The molecular classifications of phenolic compounds derived from the cinnamic and benzoic acids from *Posidonia oceanica* [6], flavonoids [7], stilbenoids [8], triterpenoids, steroids from *Ganoderma* [9], isoflavonoids from *Dalbergia parviflora* [10], sesquiterpene lactones [11, 12] and artemisinin derivatives [13] were informed. A tool for interrogation of macromolecular structure was reported [14]. Mucoadhesive polymer hyaluronan favours transdermal penetration absorption of Caff [15, 16]. Polyphenolic phytochemicals in cancer prevention, therapy, bioavailability and bioefficacy were reviewed [17]. The natural product antioxidants from herbs and spices improved the oxidative stability (OS) of vegetable oils (VOs) [18]. The VO composition–OS relationship was revealed *via* a multifactorial approach [19]. It was informed chemical and biological screening approaches to phytopharmaceuticals [20], cultural interbreeding in indigenous and scientific ethnopharmacology [21], medicinal plants (MPs) ethnobotanical studies, underutilized wild edible plants, food, medicine [22], biodiversity as a source of drugs, *Cordia*, *Echinacea*, *Tabernaemontana*, *Aloe* [23], immunomodulatory molecules from Himalayan MPs [24], antimicrobial, antioxidant, composition of *Verbena carolina*, *Mentha* [25], phylogenesis by information entropy, avian birds, influenza [26], pain, pleasure [27], ethnoscience, alternative medicine, inflammation and pain models [28]. The purpose of the present work is to review the face of pain, science, ethics of a universal symptom, hot, spicy, fresh, SA, ASA, aspirin, paracetamol, diazepam, morphine, diacetylmorphine, stimulants, adrenaline, amphetamine, Caff, THC, dopamine, 3,4-dihydroxytyrosine, cocaine, nicotine, 5-HT, fluoxetin, MDMA, advances in pain pharmacological treatment with opioid analgesics, chronic pain having no age in the 21st century, drugs in Society 4.0, this being what it is, laws and drugs. The aim of this study is to initiate a debate by suggesting a number of questions (Qs) that can arise when addressing subjects of pain, pleasure, the historical development of opioid analgesic agents, pharmacological aspects of opioid analgesics, the clinical use of opioid analgesics, pain avoidability, drugs in society and laws, in different fields and providing, when possible, answers (As), facts (Fs) and hypotheses (Hs) on pain, pleasure, advances in pain pharmacological treatment with opioid analgesics, drugs, society and laws.

## 2. Hot, Spicy and Fresh

Hot, spicy and fresh flavours are chemical stimulations of pain [29]. Two types of nerves for pain exist: Class A are thin fibres that carry the signal quickly ($20 m \cdot s^{-1}$); class C are thicker and carry the signals more slowly ($1 m \cdot s^{-1}$), which signals are called *fast* and *slow pain*, respectively. The former are the response to injuries and are usually acute and localized. The latter are usually painful and delocalized sensations. Both types enter the spine along with the nerves responsible for the sensation of temperature: They stimulate the neurones that reach the brain and their signals undergo some local process. A characteristic of pain nerves is their interaction in a gelatinous part of the spinal cord called the *substantia gelatinosa* (SG). The signals that arrive along A fibres excite SG cells but those *via* C fibres inhibit them. The net effect may be the inhibition of the cells responsible for transmitting signals A/C to the brain-processing centre (thalamus). A complex interconnection exists between the signals that initially arrive as fast and slow pain. In response to pain signals, the brain secretes its own pain relievers (*endorphins*, *enkephalins*). Both are polypeptides but endorphins present longer chains than enkephalins, which affect the transmission of nerve signals, and both are mimicked by opiates. The pain receptors that initiate the complex cascade present nerve endings that are branched: No specifically innervated pain receptors exist. However, receptors exist that respond to thermal stimulation. They are of two types: Some respond to heat and others, to cold; the latter are 10 times more numerous. Fibres A/C transmit their signals, e.g., the painful ones, to the thalamus, so that an intense thermal stimulation can be interpreted as pain. Most spices used in curries and other foods stimulate nerve endings that detect pain in the mouth (and elsewhere), but molecular structure–response relationship is unknown. *Harmful heat* (above 52°C) stimulates one receptor, moderate heat (42°C) and capsaicin, another, and the cold or freshness (below 22°C), and menthol, other different. The missing receptor in 22–24°C is in the skin, tongue, brain and nerve tissue. Its wide distribution may be because of its role in the response to injury and inflammation.

## 3. SALICYLIC ACID $C_7H_6O_3$ AND ACETYLSALICYLIC ACID $C_9H_8O_4$

Some of the softest analgesics, which are not addictive, which are in the market, are type *aspirin* [29]. They are derived from *acetylsalicylic acid* (*cf.* Figure 2), which name comes from Latin *Salix alba* or white willow.

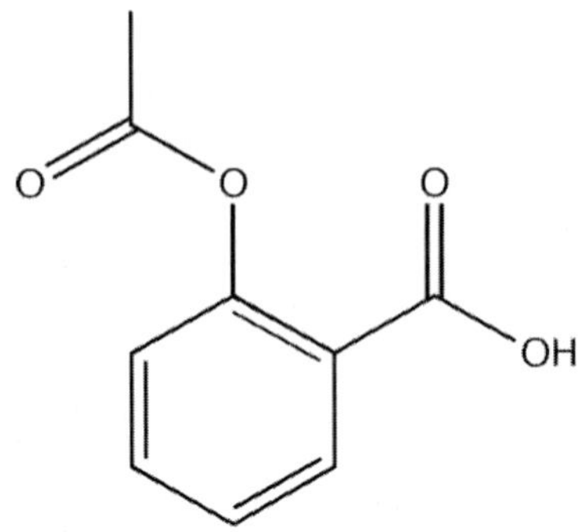


Figure 2. Molecular structure of acetylsalicylic acid (ASA, aspirin).

*Salicylic acid* (*cf.* Figure 3) is in white-willow bark, combined with a sugar molecule forming glycoside *salicin*, which was used to relieve pain in traditional medicinal (TM) preparations [29]. Leaves and bark extracts from a number of trees and shrubs, e.g., tea tree, present similar medicinal properties and contain alike substances.

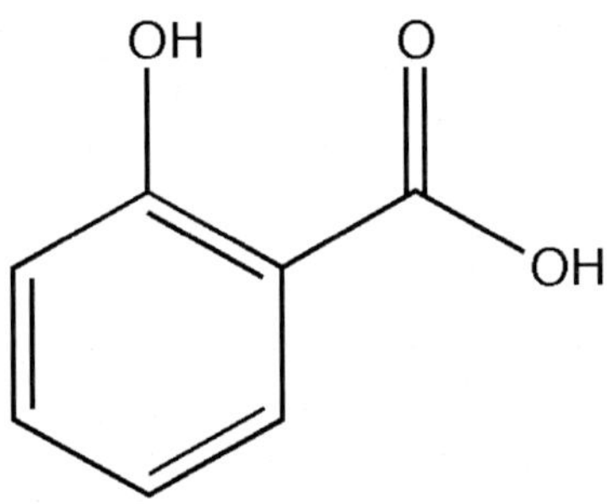


Figure 3. Molecular structure of salicylic acid (SA).

The SA, in its Na salt form, presents AEs in the stomach [29]. An improvement with regard to its tolerance was discovered by Hoffmann (1897), who worked for Bayer and was searching for a treatment for his father, who suffered from rheumatoid arthritis. The compound was marketed (1899) under the trademark *Aspirin*, which is protected in countries, e.g., Germany, but is generic in many others. Its way of action remained unknown for much time. It interferes with the synthesis of compounds [*prostaglandins* (PGs)], inhibiting the action of PG *cyclooxygenase* (COX) enzyme. The PGs are hormones that act locally and participate in many processes that happen in the body; e.g., the modification of the signals transmitted *via* synapse (connection between neurones), particularly pain signals. They intervene in blood-vessels dilation, which causes headache (if vessels are intracranial) or migraines (if vessels are outer cranium), in which cases, aspirin and other local analgesics act inhibiting the synthesis of PGs, which are those causing and transmitting pain. The zigzag lines, which persons suffering for migraine see, are caused by the constriction of the blood vessels of the brain region responsible for vision. As this presents a different cause from headache, it can be treated independently with small doses of substances, e.g., amyl nitrite, which dilate blood vessels. Aspirin applications seem to present no end. It reduces the risk of heart attacks and strokes recurrence, because it reduces the trend of platelets to gather and blood flows more freely, which was explained by Kafka to his fiancée Bauer in a not much romantic way: Aspirin is one of the few things that relieve the unbearable pain of being. Trademark *Alka-Seltzer*, popular since a number of decades, consists of aspirin, anhydrous *citric acid* (*cf.* Figure 4) and $NaHCO_3$.

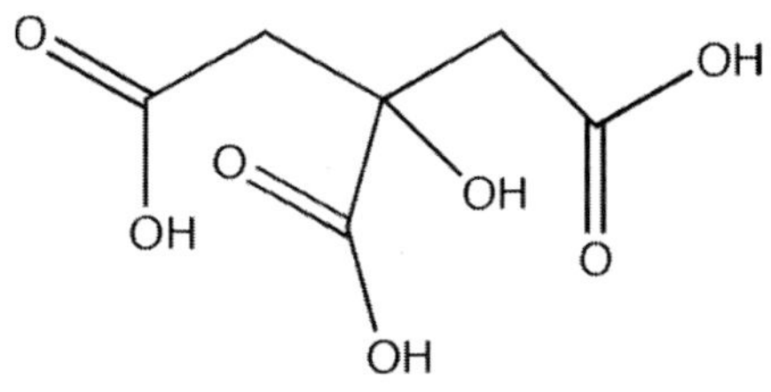


Figure 4. Molecular structure of citric acid.

When the mixture above is added to water, aspirin (an acid) becomes converted into its Na salt that is more soluble, and citric acid releases *carbon dioxide* $CO_2$ (*cf.* Figure 5) from $HCO_3^-$ ions, producing effervescence and a more pleasant taste than aspirin [29].

Figure 5. Molecular structure of carbon dioxide $CO_2$.

## 4. N-ACETYL-*para*-AMINOPHENOL $C_8H_9O_2N$

Analgesic properties of medicines, as those seen above, were discovered by accident when a certain amount of *acetanilide* (like molecule in Figure 6 but without –OH group) was added to the compound prescribed to a patient [29]. Acetanilide can, in many cases, be toxic, so that lesser harmful components were searched; e.g., N-*acetyl*-p-*aminophenol* (marketed as *Tylenol*). In fact, it becomes converted into paracetamol in the body, which generates its analgesic properties; however, part of it is converted into *aniline*, a benzene molecule with an H-atom replaced by $-NH_2$ group, which is toxic.

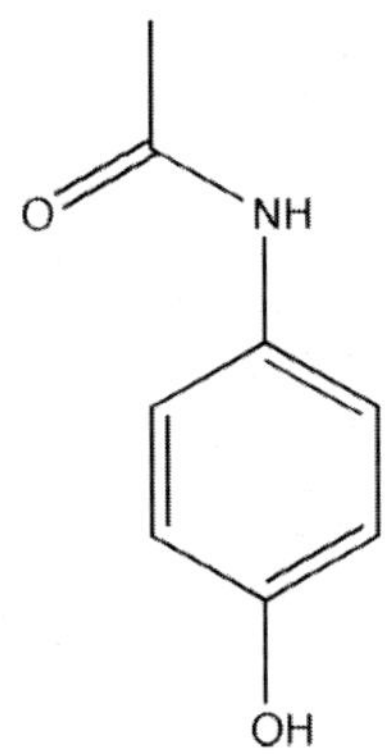


Figure 6. Molecular structure of *N*-acetyl-*p*-aminophenol (paracetamol, Tylenol).

Notice the similarity between acetanilide molecule and ASA (Figure 2) [29]. Although they form from different atoms, they present a similar conformation, so the same enzyme, that responsible for PGs synthesis so that paracetamol acts inhibiting their synthesis, recognizes both.

## 5. DIAZEPAM $C_{16}H_{13}ON_2Cl$

Sleeplessness is almost always because of anxiety, which is associated with a frenetic activity of neurones in the limbic system (LS), the brain region associated with emotions, which activity can be irradiated to brain stem, the most primitive part of the brain, and keep a state of constant and aggravated anxiety and alert [29]. Molecules called *benzodiazepines*, e.g., *diazepam* (*cf.* Figure 7), bind to proteins in zones where nerves link up (synapse zones), and favour the capacity that GABA (*cf.* Figure 8) neurotransmitter join neighbouring sites in the same molecule.

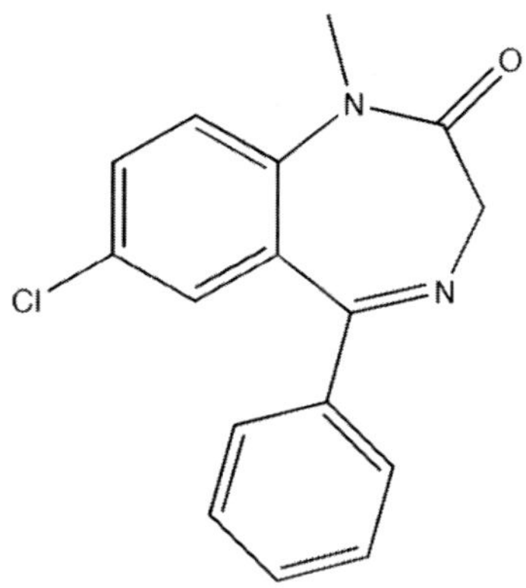


Figure 7. Molecular structure of diazepam (Valium).

The GABA inhibits nerve transmission, so that benzodiazepines promote the inhibition [29]. The sites to which benzodiazepines join are abundant in LS, so that on ingesting the compounds, the abnormal neurone activity that take people to experiment the generalized worry and fear (*anxiety*) is suppressed; i.e., benzodiazepines are medicines specific to treat anxiety that act indirectly like tranqullizers. They are used in alcohol-detoxification processes to avoid or mitigate the abstinence syndrome.

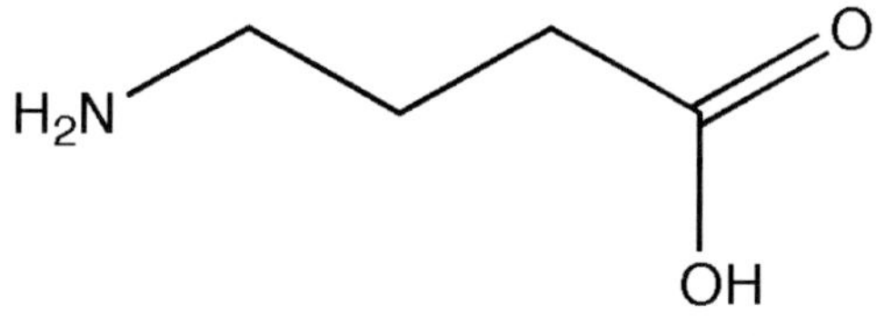


Figure 8. Molecular structure of γ-aminobutyric acid (GABA).

Benzodiazepines are slightly addictive but not lethal [29]. However, alcohol and benzodiazepines join in neighbouring sites of the same protein. Both together can change the conformation of such protein, so that GABA present a high inclination to bind it. If this happens, a massive inhibition of the nervous-system (NS) activity occurs and death can happen. Benzodiazepines include diazepam molecule (*Valium*), which is used as muscular relaxing and tranquillizer. If Cl atom on the left side is substituted by $-NO_2$, and $>NCH_3$ by $>NH$, one obtains *nitrazepam*, which trademark is *Modagon*. Another widely used tranquillizer is *chlordiazepoxide* (*Librium*).

## 6. Morphine $C_{17}H_{19}O_3N$ and Diacetylmorphine $C_{21}H_{23}O_5N$

Morphine (cf. Figure 9) is the main component of opium (from Greek opion, meaning poppy juice), which is obtained as a milky liquid exuded from the immature capsules of poppy Papaver somniferum [29]. It was isolated by Sertürner (1803–1805). Unlike aspirin, morphine and related compounds act on central NS (CNS) and induce addiction. The specific action of the compounds is related with the capacity of a molecule of fitting and blocking a specific receptor placed on the neurones.

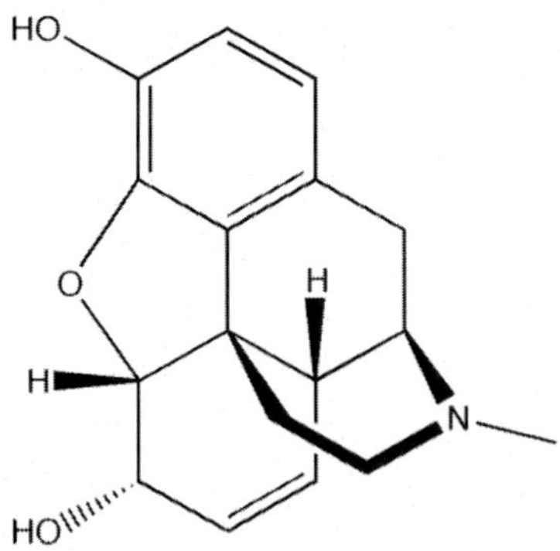


Figure 9. Molecular structure of morphine.

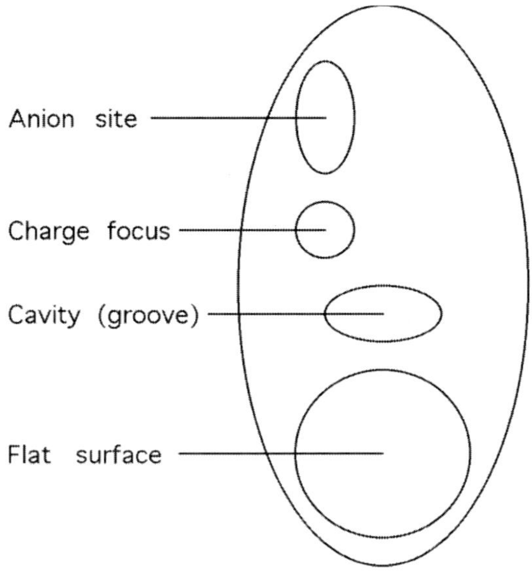


Figure 10. Tentative conformation for the protein acting as morphine and related-molecules receptor.

The proposed receptor conformation is shown in Figure 10 [29]. The benzene group of the morphine molecule fits tightly on the flat side, and the neighbouring group of C-atoms is at the correct distance and orientation to fit in the groove. Behind such groove, an anionic group exists, which attracts cationic N-atom. On fitting with receptor shape and binding it, morphine molecule removes its action: Morphine imitates body natural analgesics, enkephalins.

Morphine acts on deep or slow pain but it presents no effect on fast pain [29]. Substantia gelatinosa (SG), the spine region when pain signals are first processed, is filled with morphine receptors: There morphine acts increasing the threshold to which people feel slow pain. Morphine receptors abound in the medium part of thalamus, the brain part that acts as entry of slow-pain signals. High concentrations of the receptors exist in LS, region associated with emotion. A close relative of morphine is codeine, which is obtained when –OH group (Figure 9, left) is substituted by $-OCH_3$. In the body, $-OCH_3$ is replaced by –OH, so that codeine becomes converted into morphine. Diacethylmorphine (diamorphine, heroin, cf. Figure 11) is a morphine molecule in which H-atoms of both –OH groups were replaced by two acetyl –$COCH_3$ groups. It was synthesized by Wright (1874). The substitution of H that bonds –OH groups makes heroin lesser water soluble than morphine, but more soluble among the hydrocarbon (HC) chains of fats. Although it must be injected directly into blood, it passes much faster the blood–brain barrier (BBB, which prevents that big and water-solved molecules pass to brain). The result is that it is more powerful than morphine (more heroic) but its effect does not last so much. Once heroin became absorbed, acetyl groups are removed, which transforms it into morphine, causing an analgesic and euphoric action.

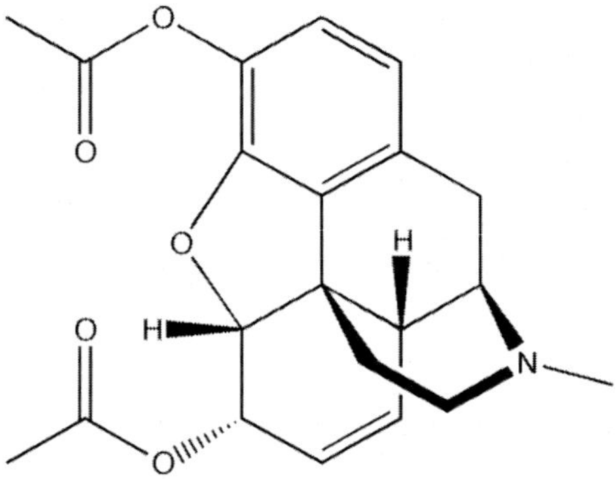


Figure 11. Molecular structure of diacethylmorphine (diamorphine, heroin).

## 7. STIMULANTS

Many drugs are not used for medical purposes, which has to do with the degree of social acceptance [29]. Some are used providing a euphoria feeling, e.g., *ethanol* $CH_3$–$CH_2$–OH (*cf.* Figure 12, a depressant).

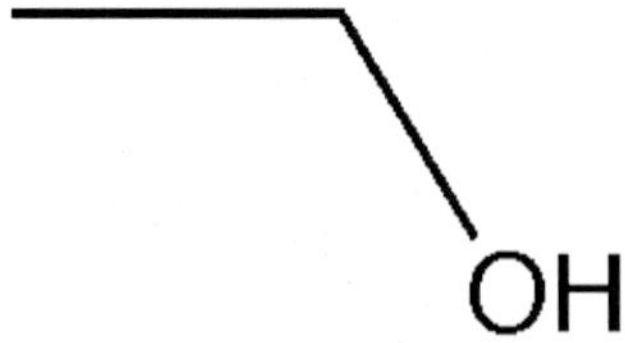


Figure 12. Molecular structure of ethanol.

The brain region associated with emotions is LS: Stimulation of LS neurones, if neurotransmitters capacity to make them communicate with each other *via* synapse is enhanced, and the capacity of inhibitory neurotransmitter GABA (Figure 8) to join them is lowered, can generate exaggerated responses of emotional activity [29]. Many connections in LS include branched neurones that originate in a small blue region of the encephalic trunk [*locus coeruleus* (LC)], which neurones characteristic is that the neurotransmitter that causes synapse is a molecule {*noradrenaline* [NA, *norepinephrine* (NE) in US]}, which is abundant in LS itself, which neurones significant neurophysiological characteristic is their branching, which result is the establishment of multiple synapses with other (polysynaptic) neurones in LS and other brain regions. So much branching and synapses may be connected to emotion, whereas lesser-branched and monosynaptic neurones would be to rational functions, which, if sure, any modification in the response of the branched neurones groups can produce modifications in emotional responses. The key for such modification must be neurotransmitter NA.

## 8. ADRENALINE $C_9H_{13}O_3N$ AND AMPHETAMINE $C_9H_{13}N$

Not all stimulants need to be administered from outside: Some, e.g., *adrenaline* (*cf.* Figure 13), are synthesized in the body itself and participate in the signal transmission from one neurone to another [29]. Adrenaline [*epinephrine* in US, from Greek (*ren* is *kidney* in Latin, *nephros*, in Greek)] is produced in the kidney glands. It circulates *via* blood and affects the autonomous NS (ANS, nerves on which one has not voluntary control, which are different from CNS). It acts on heart muscles to increase the contraction force, dilates the pupils and stimulates sweat and saliva secretion. However, H-bonds between –OH and >NH groups in the molecule cause that it do not pass BBB.

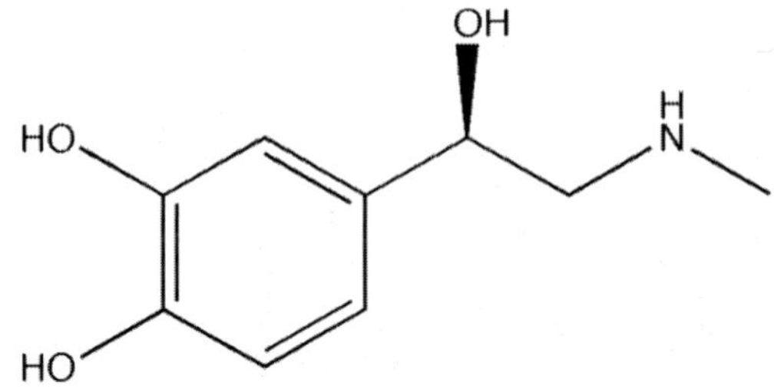


Figure 13. Molecular structure of adrenaline (epinephrine).

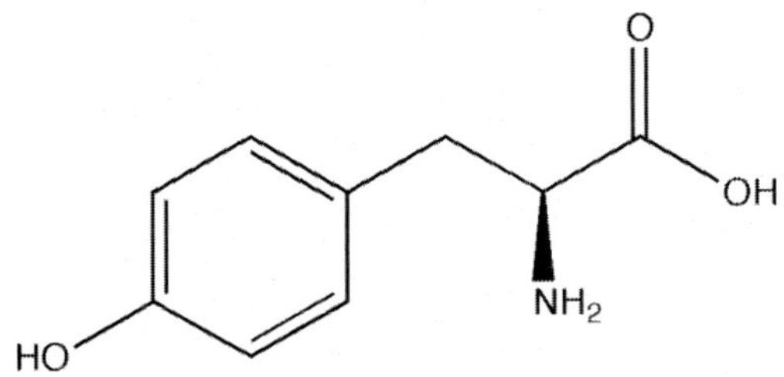


Figure 14. Molecular structure of (L)-tyrosine (Tyr, Y).

However, the brain synthesizes adrenaline and similar compounds, e.g., NA [29]. Its precursor is amino acid (AA) *tyrosine* (Tyr, Y, *cf.* Figure, 14), with the same structural planning than adrenaline, and becomes converted into NA by different enzymes. The conversion occurs abundantly in LC and LS neurones and, in any way, feeds people's emotions.

*Amphetamine* (*cf.* Figure 15) molecule was marketed under trademark *Benzedrine* as a stimulant [29]. Conformational similarity with NA suggests that it can act similarly, stimulating LS in LC.

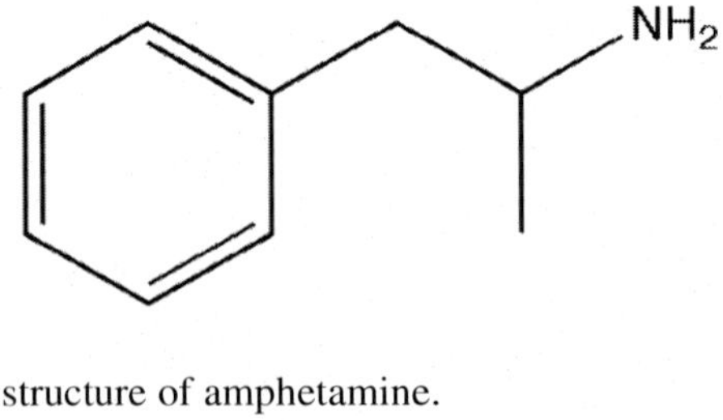


Figure 15. Molecular structure of amphetamine.

As LC is connected to upper centres (mainly brain cortex), amphetamine can stimulate cognitive functions, causing a state of higher alert and euphoria feeling [29]. Amphetamine action is not simply to play NA role as neurotransmitter. Their molecules can imitate so perfectly NA ones that they can occupy their place in storage areas of presynaptic (front) neurones, which results in NA being displaced from its storage site, entering the space between the pre and postsynaptic (posterior) neurones. On rising their concentration in the place, NA molecules join proteins on the wall of the postsynaptic neurone, changing their conformation and shooting a signal. Neuronal activity is enhanced and a euphoria, alert and power sensation occurs. Amphetamine molecules exist in two forms, one is the mirror image of the other [they are chiral compounds, e.g., *lactic acid* (*cf.* Figure 16)]. The form that rotates polarized-light plane to the right (*Dexedrine*) is much more potent than its mirror image. The two differ in shape, e.g., hands, and only one of them will fit in the glove that is the protein.

*Methamphetamine* differs from amphetamine molecule in that it presents a $-CH_3$ group instead of one H-atom of $-NH_2$ group [29]. The stimulant molecule is marketed as *Methedrine* (*speed*).

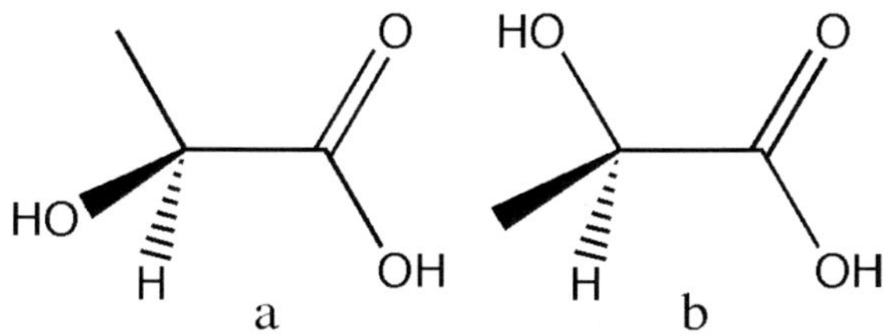


Figure 16. Molecular structure of lactic acid: (a) (L)-lactic acid and (b) (D)-lactic acid.

## 9. Caffeine $C_8H_{10}O_2N_4$

Caffeine (cf. Figure 17) is the component of coffee and tea that stimulates brain cortex, inhibiting the phosphodiesterase enzyme that, in turn, degrades a 2$^{nd}$ messenger from the same family as adenosine triphosphate (ATP, cf. Figure 18) [cyclic adenosine momophosphate (cAMP)] [29].

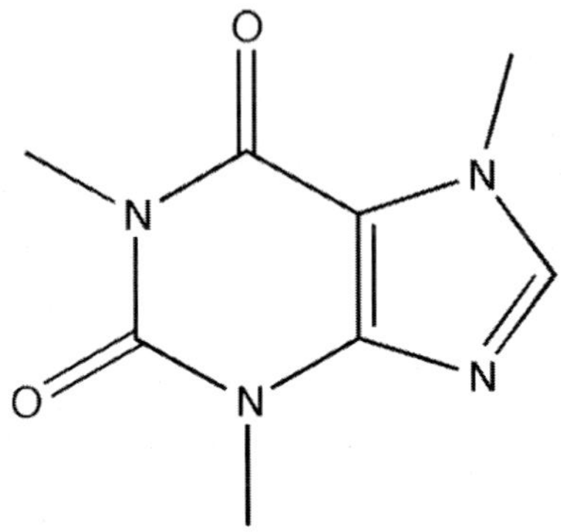


Figure 17. Molecular structure of caffeine (Caff).

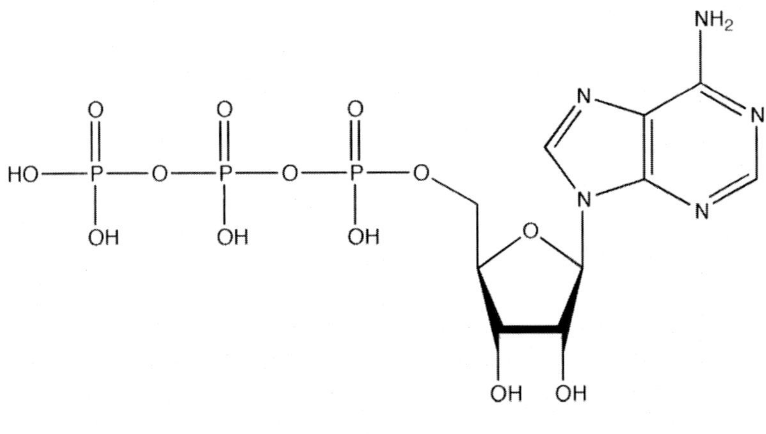


Figure 18. Molecular structure of adenosine triphosphate (ATP).

Molecule of Caff could be taken for ATP: A comparison of the structures of both molecules shows that they include two fused heterocyclic rings [29].

Typically, a cup of coffee or tea contains 1/10 of Caff [29]. Coffee is obtained from roasting the seeds of tree *Coffea arabica*, and tea, from the fermented leaves of *Camellia thea*. Caffeine occurs in the seeds of cola plant

*Cola acuminata* and *C. nitida*, which grows in East Africa and South America. The plants extracts are used to spice (and by their stimulant action) cola drinks instead of *cocaine*, which is what they originally contained. *Theobromine* is a Caff close relative, which molecule differs in that it presents an >$NCH_3$ group substituted by >NH. It is the stimulant that occurs in chocolate. Black chocolate is cacao and sugar. Milk chocolate contains additional ingredients, e.g., vanilla cream, etc. Theobromine provides a stimulating effect lesser than Caff.

## 10. TETRAHYDROCANNABINOL $C_{21}H_{30}O_2$

Tetrahydrocannabinol (cf. Figure 19) is the active component of cannabis, which is obtained from plant Cannabis sativa, from the same family as hop, plant used to make beer [29]. Cannabis was cultivated for its fibres (hemp), which are used in industry to make sailcloth, clothes and ropes, which derives from Latin cannabis. The female plant exudes a resin that covers near flowers and leaves, and such resin is added to preparations that humans consume in different forms. Marijuana is obtained from dried and ground leaves and flowers; hashish is the treated resin of cannabis. The way of action of the high-structural-complexity molecule was not determined.

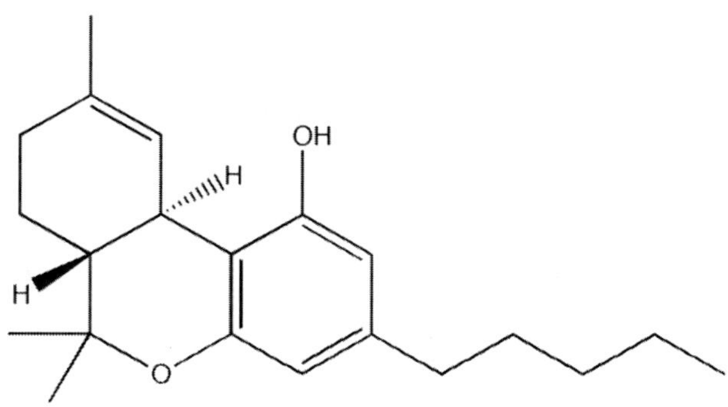


Figure 19. Molecular structure of tetrahydrocannabinol (THC).

## 11. DOPAMINE $C_8H_{11}O_2N$ AND 3,4-DIHYDROXYTYROSINE $C_9H_{11}O_4N$

*Dopamine* (*cf.* Figure 20) is an important neurotransmitter [29]. Its distribution in brain is not at all uniform as it usually concentrated in *corpus striatum*, which is the brain region involved in movement co-ordination, and LS. A dopamine-production deficit is responsible for causing Parkinson's disease (PD), whose main symptoms are rigidity, trembling and hypokinesia (slow to start movements). A dopamine excess in the brain is related to the beginning of Huntington's disease (HD, chorea), genetic upset, which symptoms are marked involuntary movements. It is involved, as neurotransmitter, in the brain routes that control reinforcement and pleasure, and addictive processes (e.g., cocaine, *nicotine*).

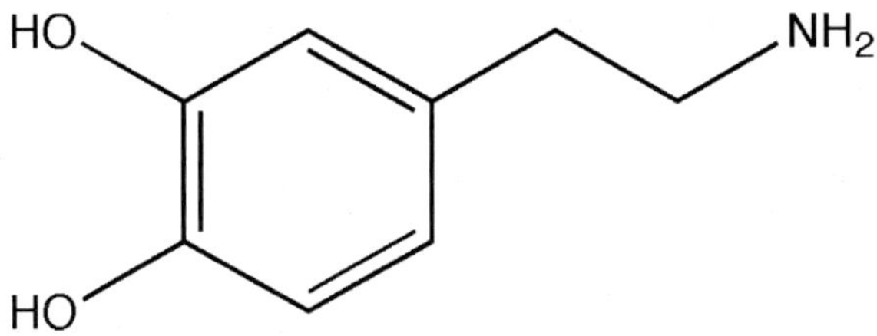


Figure 20. Molecular structure of dopamine.

Figure 21 shows *(L)-3,4-dihydroxyphenylalanine* [*levodopa*, (*L*)-DOPA, *3-hydroxytyrosine*] molecule: *L* comes from the chiral form (*levo*, left) in which it occurs [29].

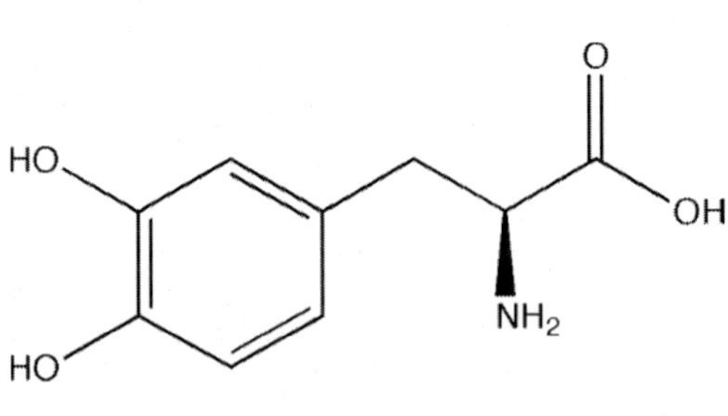


Figure 21. (L)-3,4-dihydroxyphenylalanine [levodopa, (L)-DOPA, (L)-3-hydroxytyrosine].

Dopamine is synthesized *via* (*L*)-DOPA from Tyr (Figure 14) in *substantia nigra* (SN), which colour is because of *melanin* (*cf.* Figure 22) accumulation, which is made up of Tyr [29]. As SN cells die, dopamine production decays and symptoms intensify. An HD-symptoms temporal relief is gotten administering (*L*)-DOPA but cells continue dying. Dopamine itself does not pass BBB so it is ineffective when it is administered intravenous (i.v.), but 1% of (*L*)-DOPA opens pass to the brain when it is ingested and, once there, it becomes converted into dopamine by –COOH-group loss. As amount is small, high and frequent doses are needed but its effectiveness decays as HD advances.

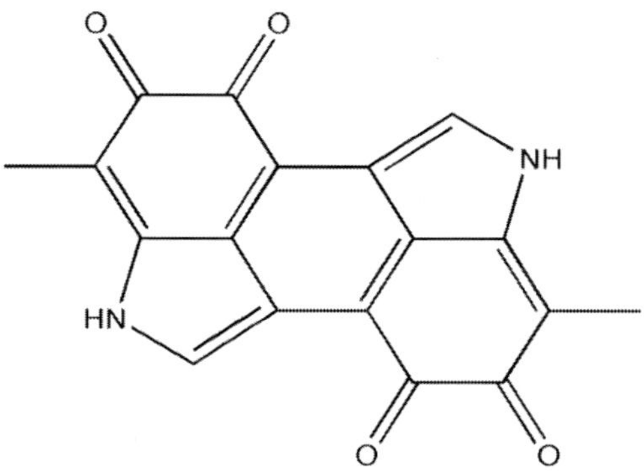


Figure 22. Molecular structure of melanin.

Symptoms of PD can be caused by illegal-heroin consumption, sometimes contaminated with compounds that kill SN cells and cause PD [29].

## 12. COCAINE $C_{17}H_{21}O_4N$

Cocaine (coke, cf. Figure 23) is an alkaloid that is obtained from the leaves of plant coca Erythroxylon coca, endemic in the hills of East Andes and widely cultivated in Bolivia, Peru, Ecuador and Colombia [29]. Niemann isolated it (1860). Coca developed cocaine production as pesticide, as it inhibits neurotransmitter-octopamine reuptake, specific of insects. In humans, it inhibits dopamine, NA and serotonin reuptake in the brain. Such neurotransmitters are in the synaptic zone for a longer time: Cocaine stimulates the release of more dopamine. The immediate effect is euphoria, verbosity and motor activity but the effects pass at once and are succeeded by a depression and anguish feeling, which only calms down consuming more.

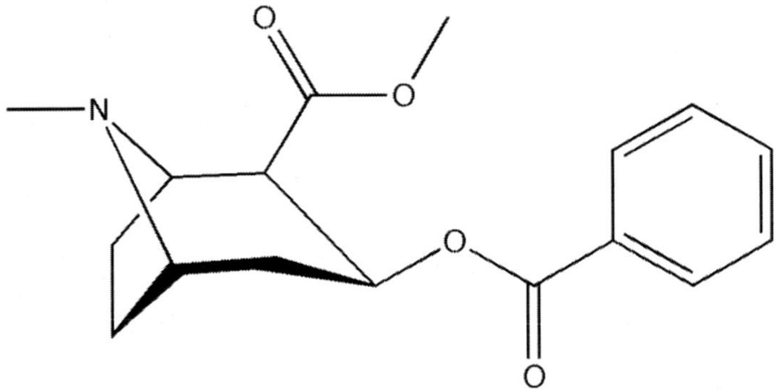


Figure 23. Molecular structure of cocaine (coke).

The habitual form of administering cocaine is as its chloride (salt formed on combining cocaine with HCl) by nasal inhalation; however, the result makes that the nose capillary vessels constrict, which can cause that the nasal septum necrose [29]. Crack is cocaine to which HCl was removed, heating the compound in a $NaHCO_3$ solution until water evaporate. The free base that is obtained makes a sound like crack when heated because of $CO_2$ release from $HCO_3^-$ (here the name). As the free base is more molecular than ionic (not so in $Cl^-$), it evaporates at low temperature and can be inhaled via a heated tube. It was used for eye and mouth operations [later, it was substituted by local anaesthetics (e.g., lidocaine, novocaine)], and oral and throat pains. It has no medical uses but as soft anaesthetic in ophthalmology.

## 13. NICOTINE $C_{10}H_{14}N_2$

Nicotine (cf. Figure 24) is found in the leaves of plant tobacco Nicotiana tabacum (called by one of the first tobacco importers, French ambassador in

Portugal Nicot) in combination with citric (Figure 4) and malic (cf. Figure 25) acids [29].

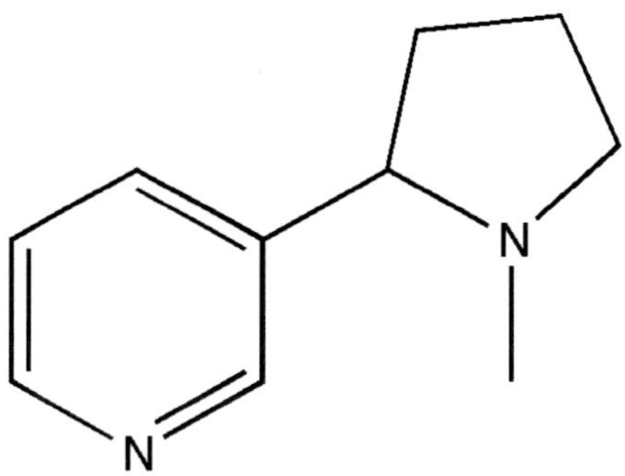


Figure 24. Molecular structure of nicotine.

Alkaloids occur frequently in plants to defend themselves *vs.* predators, because of its unpleasant taste [29]. When tobacco is harvested, leaves dry, which causes that proteins, carbohydrates ($CH_2O$) and alkaloids composition change. For cigarettes tobacco, leaves are hung in a room with ventilation and a heat source. After a week, green leaves turn yellow as carotenes are made dominant. Leaves present greater sugars and nicotine concentration. Cigars tobacco is cured in the air, hanging them for a number of weeks. The resultant product is low in sugars and variable in nicotine. Tobacco curing at fire causes that leaves green be exposed to wood smoke and heat, and is used to smoke a pipe: It is low in sugars and nicotine. After curing, leaves are fermented storing them in a humid place: Taste changes on rehydrating, which makes them lesser fragile.

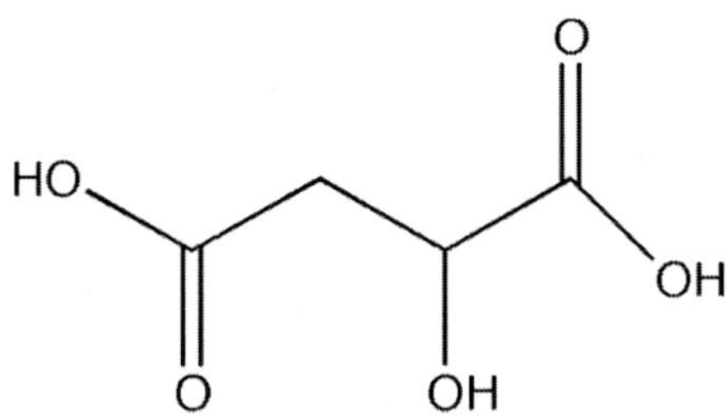


Figure 25. Molecular structure of malic acid.

A typical cigarette contains 10mg nicotine, 1mg of which is absorbed by the smoker, mainly *via* the lungs, where it arrives transported by tar molecules [29]. Pipe and cigar tobacco smoke is lesser acidic than cigarettes, in such a way that nicotine in both is more similar to its basic form. The result is that more nicotine becomes absorbed in the mouth and nasal cavity. Cigarettes cause that nicotine concentration in blood rise faster than with cigars or pipe consumption, which causes a slower but lasting concentration. Nicotine presents complex and numerous pharmacological effects, some of which are identical to cocaine (Figure 23), and final result depends on dose. Smokers experience sometimes alert feeling and, others, relaxation. Nicotine, like cocaine, heroin and marijuana, increases dopamine (Figure 20) level, which affects the routes by which brain controls pleasure. As it causes an adrenaline unloading from the adrenal cortex, its ingestion causes an immediate *kick*. Adrenaline, *via* the endocrine system, stimulates glucose release and alert feeling, which stimulation is followed by depression and fatigue. Hormone corticosterone, which is released in a stress episode, reduces nicotine effects, which causes that more amount be required to get the same effect. Nicotine is addictive and smokers use to reject the cigarettes that do not present it. Cigarettes with low tar content present a low substance in nicotine but with these, smokers inhale more strongly and smoke with more frequency: As a result, the net effect of passing to low-tar cigarettes is a rise in CO enter rate, with the underlying risk of suffering cardiovascular (CV) accidents. Menthol cigarettes increase nicotine addiction. In some countries, e.g., Pakistan, clove is added to cigarettes.

## 14. Serotonin $C_{10}H_{12}ON_2$ and Fluoxetin $C_{17}H_{18}ONF_3$

Serotonin (5-hydroxytryptamine, cf. Figure 26) is a neurotransmitter and a slightly modified version of essential AA tryptophan (Trp, W) used in proteins manufacturing [29]. Although food contains 5-HT, most part forms in the body by enzyme action on Trp. The 5-HT participates in a wide number of activities, especially in CNS, modulating courage, emotion, sleep and appetite. The cerebrospinal fluid of the persons that suffer for depression contains lower concentrations of substances derived from 5-HT, and the same happens in suicidal cases, which low concentrations can be because of an abnormally slow Trp transport via BBB.

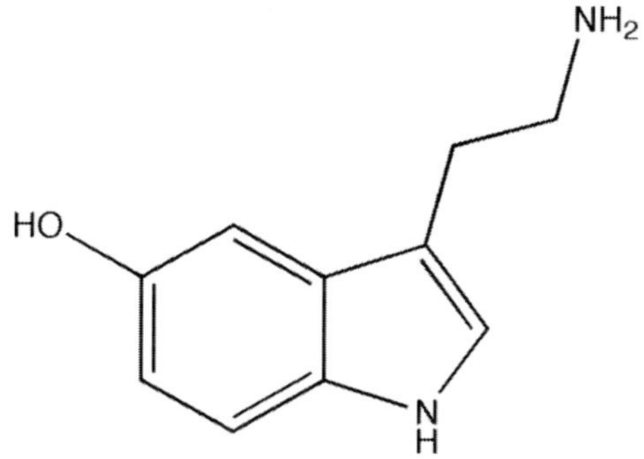


Figure 26. Molecular structure of serotonin (5-hydroxytryptamine, 5-HT).

The 5-HT is a powerful vasoconstrictor [29]. When it acts on veins smooth musculature, it hinders capillaries blood flow and causes blush. Certain hallucinogen drugs affect its mechanism of action. The most famous is lysergic acid diethylamide (LSD, trippy) but its precise (probably multifaceted) action mode remains unknown. Melatonin is closely related with 5-HT: benzene –OH group is substituted by $-OCH_3$, and side-chain $-NH_2$ becomes converted into $-NHCOCH_3$. Melatonin is produced in the pineal gland, a structure remembering a 3rd eye that Descartes thought that it was where the soul rested, in the vascular side of BBB. It is believed that melatonin controls hormones amounts during the day and it has to do with jet lag. Sun stimulates the pineal gland to free 5-HT and melatonin, although an excess of the latter causes dullness, lethargy and sleepiness. Seasonal affective disorder (SAD) or winter depression occurs when persons are exposed to too low levels of sunlight for long periods, e.g., in countries of high latitudes. Nobody really knows SAD cause but a number of hypotheses exist. A possible cause results that low light levels be not enough to produce melatonin or 5-HT suitable amounts. Fluoxetine (cf. Figure 27) combining with HCl forms the fluorinated salt marketed as Prozac (Sarafem), an antidepressant. In a depression, a 5-HT disequilibrium exists in synapse; e.g., not much 5-HT is present to stimulate synapse. Prozak inhibits 5-HT reuptake, i.e., avoids that 5-HT molecules that exist in synapse return to the presynaptic neurone. The result is that 5-HT is at greater concentration more time in the synapse site and could join the postsynaptic neurone, which replies sending the signal, e.g., in a non-depressed individual. Prozac is specific for 5-HT receptors and presents lesser secondary effects than the drugs that interact more generally.

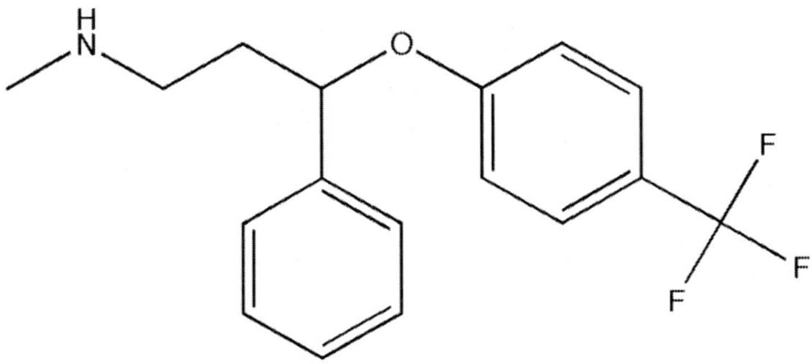


Figure 27. Molecular structure of fluoxetine (Prozac, Sarafem).

## 15. Methylenedioxymethamphetamine $C_{12}H_{17}O_2N$

The little monster *methylenedioxymethamphetamine* (Figure 28) exists in the street as *ecstasy* (E, to be out of oneself, *molly*) [29]. It was synthesized by Merck Laboratories (1912) searching for drugs for the treatment of nasal haemorrhages. Its chemical structure is similar to *methamphetamine*, *methylenedioxyamphetamine* (MDA) and *mescaline*, all causing brain damage. It affects the neurones that use 5-HT and influences courage, aggressiveness, sexual activity, sleep and pain sensitivity. It developed as treatment *vs.* obesity. When it is taken for a long period (6–7 years), 5-HT receptors become exhausted. Concentration and cognitive sense are lost. In high doses, it produces a rise in body temperature, which causes that muscles, kidneys and CV system fail. At the point, one is permanently outside of him. It was used for the treatment of post-traumatic stress, e.g., cases of women that were raped.

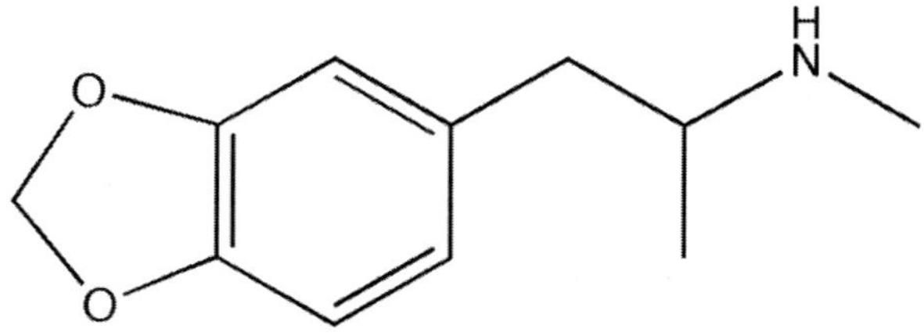


Figure 28. Molecular structure of methylenedioxymethamphetamine [MDMA, ecstasy (E), molly].

# 16. Advances in Pain Pharmacological Treatment with Opioid Analgesics

Academy of Pharmacy of Comunitat Valenciana–Aristo organized a colloquium on advances in pain pharmacological treatment with opioid analgesiscs. López Muñoz proposed Q/A/F/H on *victo dolore* and historical development of opioid analgesic agents [30].

Q1. Pain?

A1. Pain is a symptom as old as humanity itself.

F1. Sertürner (1805) discovered *morphine*.

F2. Wright (1874) synthesized *diamorphine* (*heroin*).

H1. (IASP). *Pain*: an unpleasant sensory/emotional experience associated with actual/potential tissue damage, or described in terms of such damage.

Selva Otaolaurruchi proposed H/Q/A on opioid-analgesic pharmacological aspects from pharmacist's perspective [31].

H2. Four types of pain exist: cognitive, physiological (opiate related), social and emotional pains.

Q2. How to choose the opioid?

Q3. What is the action they have?

A3. Analgesic action.

H3. Three types of pain receptors: μ-, δ- and κ-receptors.

H4. Treatment: opioid rotation.

López Alarcón proposed Q/A considering opioid-analgesic clinical use/responsible management [32].

Q4. Electronic (e)-prescription, how do we prescribe?

Q5. How do we prescribe at home?

Q6. What to watch?

A6. The four As of watching: Analgesic, Activity, Aberrant behaviour and AEs.

## 17. CHRONIC PAIN HAS NO AGE IN THE 21ST CENTURY

Grisolía and de Andrés organized Day *Chronic Pain Has No Age in 21st Century* [33]. Borrás Blasco proposed H/Q on ageing and changes in pain perception [34].

H1. *Ageing*: process during which structural/functional changes accumulate in an organism as a result of passing of time.

Q1. How does the perception of pain change with ageing?

H2. Ageing leads to *decay* in *adaptive capacity*.

H3. Pain door theory.

H4. Referred pain.

H5. Pain → Fragility → Disability.

She provided a conclusion (C).

C1. The predominant pain in the elderly is chronic pain.

Mínguez Martí proposed H on polypharmacy, interactions and limitations for pharmacological prescription in chronic pain [35].

H6. Serotonin syndrome: Tramadol/Meperidine/Fentanyl increase 5-HT levels if co-administered with others modulating 5-HT route.

H7. Keep paracetamol as *opioids saving*.

De Andrés proposed H on alternatives in supervising treatment, endoscopic and minimally invasive techniques [36].

H8. (Grisolía). Pain is avoidable.

H9. Pain is avoidable because pain is treatable.

H10. Myofascial pain syndrome: Pressure on sensitive points in your muscles causes pain in muscle and sometimes in seemingly unrelated parts of your body.

He provided a conclusion.

C2. Change of paradigm in pain treatment.

Raya raised Q on illuminating confusion, stem cells (SCs)' riot, scientific/terminological clarification on altars of its correct application in present/future medicine [37].

Q2. The SCs: cancer predisposition?

Blanquer raised the following questions on MMSCs [38].

Q3. The MMSCs, where are they obtained?

Q4. The MMSCs, which are their applications?

Q5. Pain related to exercise: The new fashions: running, fitness… What exercise to do?

Q6. Pain related to exercise: The new fashions: running, fitness… How to prepare oneself?

Q7. Where do people talk when talking about fat?

He provided a conclusion.

C3. The MMSCs are advanced therapy drugs.

Moraleda proposed H/Q on blood as regenerative-therapies origin and applications in body improvement/pain control [39].

H11. Change of paradigm: We do not work with medicines but cells.
Q8. Relapse: antitumour immunotherapy?
Q9. Chimeric antigen receptor (CAR)-T cells: How do they work?
H12. (TerCel, 2019). *White Book of Cell Therapy in Spain* [40].

## 18. DRUGS IN SOCIETY 4.0

Group SOLINDRUGS organized Workshop *Drugs in Society 4.0* [41]. Ventura proposed H/Q on substances analysis as tool to monitor drug recreational market, use patterns/detected trends [42].

H1. Attractors of *false ecstasy*.
H2. (Giné et al., 2013). New psychoactive substances (NPSs) adulterating controlled drugs [43].
Q1. The NPSs as adulterants of controlled drugs: A worrying phenomenon?
Q2. What have they found?
Q3. What are the implications of these findings?

Pérez proposed questions/answers on current challenges in drug control labratories [44].

Q4. How to monitor the drug recreational market?
Q5. How does people know the appearance of NPSs?
Q6. When do NPSs begin to be detected?
Q7. Economic responsibilities?

Q8. What do people know about their origin?

A8. China and in bulk.

Q9. How have European countries reacted?

Q10. What do people know about synthetic cannabinoid agonists?

Q11. When are they detected in the illegal drug market?

Q12. How are they called?

Q13. How do people differentiate the new ones?

A13. Method European Monitoring Centre for Drugs and Drug Addiction (EMCDDA).

Q14. What is its toxicity?

Q15. Synthetic catinones, when are they first detected?

De San Román provided conclusions on Ai Laket!!, a story for risk reduction [45].

C1. It is essential to be at street level doing prevention.

C2. One should adapt himself to leisure and black market.

C3. The importance of gender.

Bravo proposed H on controlling drug precursors [46].

H3. (Untrue H). Oversight bodies invent stories.

Hernández proposed H on drug use from a global analytical chemistry perspective/wastewater-analysis contributions [47].

H4. Wastewater-based epidemiology.

Gallart proposed the following hypothesis on contributions to drug analysis [48].

H5. Ion mobility spectrometry (IMS).

Costero proposed hypothesis on drugs, colours and protection *vs.* chemical submission [49].

H6. γ-Hydroxybutyric (GHB) acid (*cf.* Figure 29).

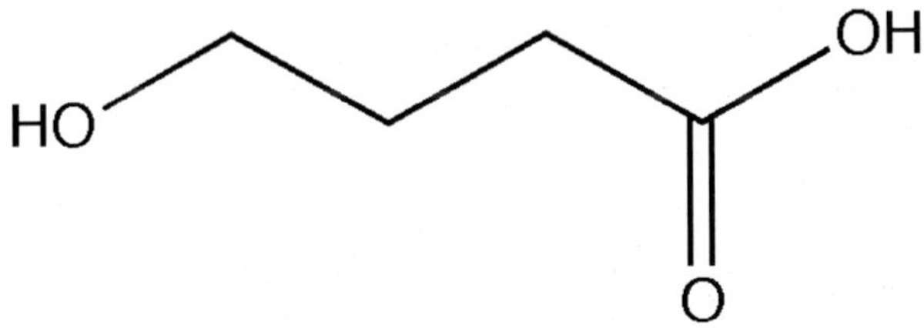

Figure 29. Molecular structure of γ-hydroxybutyric (GHB) acid.

Sorribes proposed a hypothesis on drug analysis in biological fluids *via* IMS [50].

H7. IMS problem: low sensibility in complex samples. e.g., saliva.

## 19. This Is What It Is: Laws and Drugs

Consumo-ConCiencia proposed questions and answers on this being what it is, laws and drugs [51].

Q1. Why is it a crime to sell drugs?

A1. Drugs are harmful to health and trafficking with them means endangering the health of others.

Q2. Are alcohol, coffee, tobacco or psychoactive drugs not harmful to health or are they not drugs?

A2.Yes, of course they are drugs and can be harmful to health.

Q3. In addition, why are they not prohibited like the others?

A3. Some drugs are legal and some others are not *for economic-political and cultural reasons*.

Q4. What drugs are illegal?

A4. Cannabis, MDMA (*ecstasy*), cocaine, heroin, LSD (*trippy*), psilocybin (*magic mushroom*), 2,5-dimethoxy-4-bromophenethylamine (2C-B, *nexus*), *etc*.

Q5. Is using illegal drugs a crime?

A5. Consuming any substance *in a private space* is not a crime, but any other situation, yes.

Q6. What if one is a minor?

A6. If one is a minor Criminal Code would not apply but a specific law regulating minors.

Q7. Is it legal to grow marijuana at home?

A7. One can have a *crop for self-consumption in his own home if it is not visible from outside*.

Q8. What is the *drugtest*?

A8. A device used in traffic controls that detects *via* saliva if drugs were consumed in last hours.

Q9. What happens if one tests positive for non-alcoholic drug traffic control?

A9. If one is positive, a second test will be performed of saliva but in a laboratory chromatograph.

Q10. If I refuse to take the test?

A10. If one refuses, the criminal code will apply.

## 20. Final Remarks

From the present results and discussion, the following final remarks can be drawn.

1) (International Association for the Study of Pain). *Pain*: an unpleasant sensory and emotional experience associated with actual or potential tissue damage, or described in terms of such damage.
2) Four types of pain: cognitive, physiological (opiate related), social and emotional pains.
3) There are three types of pain receptors: μ-, δ- and κ-receptors.

4) Treatment: opioid rotation.
5) The four As of watching: Analgesic, Activity, Aberrant behaviour and Adverse effects.
6) Pain is avoidable: Chronic pain has no age in the 21st century.
7) The predominant pain in the elderly is chronic pain.
8) Change of paradigm in pain treatment.
9) Multipotent mesenchymal stromal cells are advanced therapy drugs.
10) Drug-checking services and European Union's early-warning system can play an important role in reducing the harm associated with adulteration in the drug illegal market.
11) It is essential to be at street level doing prevention.
12) One should adapt himself to leisure and black market.
13) The importance of gender.
14) Some drugs are legal and some others are not *for economic-political and cultural reasons.*

## ACKNOWLEDGMENTS

The authors thank support from Fundación Universidad Católica de Valencia *San Vicente Mártir* (Project No. 2019-217-001UCV).

## REFERENCES

[1] Torrens-Zaragozá, F. Molecular categorization of yams by principal component and cluster analyses. *Nereis* 2013, 2013(5), 41-51.
[2] Torrens-Zaragozá, F. Classification of lactic acid bacteria against cytokine immune modulation. *Nereis* 2014, 2014(6), 27-37.
[3] Torrens-Zaragozá, F. Classification of fruits proximate and mineral content: Principal component, cluster, meta-analyses. *Nereis* 2015, 2015(7), 39-50.

[4] Torrens-Zaragozá, F. Classification of food spices by proximate content: Principal component, cluster, meta-analyses, *Nereis* 2016, 2016(8), 23-33.

[5] Torrens, F.; Castellano, G. From Asia to Mediterranean: Soya bean, Spanish legumes and commercial *soya bean* principal component, cluster and meta-analyses. *J. Nutr. Food Sci.* 2014, 4(5), 98-98.

[6] Castellano, G.; Tena, J.; Torrens, F. Classification of polyphenolic compounds by chemical structural indicators and its relation to antioxidant properties of *Posidonia oceanica* (L.) Delile. *MATCH Commun. Math. Comput. Chem.* 2012, 67, 231-250.

[7] Castellano, G.; González-Santander, J.L.; Lara, A.; Torrens, F. Classification of flavonoid compounds by using entropy of information theory. *Phytochemistry* 2013, 93, 182-191.

[8] Castellano, G.; Lara, A.; Torrens, F. Classification of stilbenoid compounds by entropy of artificial intelligence. *Phytochemistry* 2014, 97, 62-69.

[9] Castellano, G.; Torrens, F. Information entropy-based classification of triterpenoids and steroids from *Ganoderma*. *Phytochemistry* 2015, 116, 305-313.

[10] Castellano, G.; Torrens, F. Quantitative structure–antioxidant activity models of isoflavonoids: A theoretical study. *Int. J. Mol. Sci.* 2015, 16, 12891-12906.

[11] Castellano, G.; Redondo, L.; Torrens, F. QSAR of natural sesquiterpene lactones as inhibitors of Myb-dependent gene expression. *Curr. Top. Med. Chem.* 2017, 17, 3256-3268.

[12] Torrens, F.; Castellano, G. Structure–activity relationships of cytotoxic lactones as inhibitors and mechanisms of action. *Curr. Drug Discov. Technol.*, in press.

[13] Torrens, F.; Redondo, L.; Castellano, G. Artemisinin: Tentative mechanism of action and resistance. *Pharmaceuticals* 2017, 10, 20–4-4.

[14] Torrens, F.; Castellano, G. A tool for interrogation of macromolecular structure. *J. Mater. Sci. Eng. B* 2014, 4(2), 55-63.

[15] Torrens, F.; Castellano, G. Mucoadhesive polymer hyaluronan as biodegradable cationic/zwitterionic-drug delivery vehicle. *ADMET DMPK* 2014, 2, 235-247.

[16] Torrens, F.; Castellano, G. Computational study of nanosized drug delivery from *cyclo*dextrins, crown ethers and hyaluronan in pharmaceutical formulations. *Curr. Top. Med. Chem.* 2015, 15,1901-1913.

[17] Estrela, J.M.; Mena, S.; Obrador, E.; Benlloch, M.; Castellano, G.; Salvador, R.; Dellinger, R.W. Polyphenolic phytochemicals in cancer prevention and therapy: Bioavailability *versus* bioefficacy. *J. Med. Chem.* 2017, 60, 9413-9436.

[18] Redondo-Cuevas, L.; Castellano, G.; Raikos, V. Natural antioxidants from herbs and spices improve the oxidative stability and frying performance of vegetable oils. *Int. J. Food Sci. Technol.* 2017, 52, 2422-2428.

[19] Redondo-Cuevas, L.; Castellano, G.; Torrens, F.; Raikos, V. Revealing the relationship between vegetable oil composition and oxidative stability: A multifactorial approach. *J. Food Compos. Anal.* 2018, 66, 221-229.

[20] Torrens, F.; Castellano, G. Chemical/Biological Screening Approaches to Phytopharmaceuticals. In *Research Methods and Applications in Chemical and Biological Engineering*; Pourhashemi, A., Deka, S. C., Haghi, A. K., Eds.; Apple Academic–CRC: Waretown, NJ, in press.

[21] Torrens, F.; Castellano, G. Cultural Interbreeding in Indigenous/Scientific Ethnopharmacology. In *Research Methods and Applications in Chemical and Biological Engineering*; Pourhashemi, A., Deka, S. C., Haghi, A. K., Eds.; Apple Academic–CRC: Waretown, NJ, in press.

[22] Torrens, F.; Castellano, G. Ethnobotanical Studies of Medicinal Plants: Underutilized Wild Edible Plants, Food and Medicine. In *Innovations in Physical Chemistry*; Haghi, A. K., Ed.; Apple Academic–CRC: Waretown, NJ, in press.

[23] Torrens, F.; Castellano, G. Biodiversity as a source of drugs: Cordia, Echinacea, Tabernaemontana and Aloe. In *Green Chemistry and Biodiversity: Principles, Techniques, and Correlations*; Aguilar, C. N., Ameta, S. C., Haghi, A. K., Eds.; Apple Academic–CRC: Waretown, NJ, in press.

[24] Torrens, F.; Castellano, G. Immunomodulatory Molecules from Himalayan Medicinal Plants. In *Chemistry and Chemical Engineering for Sustainable Development: Best Practices and Research Directions*; Esteso, M. A., Ribeiro. A. C. F., Haghi, A. K., Eds.; Apple Academic–CRC: Waretown, NJ, in press.

[25] Torrens, F.; Castellano, G. Antimicrobial, Antioxidant, and Composition of Verbena carolina and Mentha. In *Applied Pharmaceutical Science and Microbiology: Novel Green Chemistry Methods and Natural Products*; Mahapatra, D.K., Talele, S.G., Haghi, A. K., Eds.; Apple Academic–CRC: Waretown, NJ, in press.

[26] Torrens, F.; Castellano, G. *Phylogenesis* by Information Entropy: Avian Birds and 1918 Influenza Virus. In *Modelling & Simulation*; Turner, S., Yunus, J., Eds.; IEEE: London, 2008; Vol 2; pp 1-2.

[27] Torrens, F.; Castellano, G. Pain and Pleasure. In *Molecular Chemistry and Biomolecular Engineering: Integrating Theory and Research with Practice*; Pogliani, L., Torrens, F., Haghi, A. K., Eds.; Apple Academic–CRC: Waretown, NJ, in press.

[28] Torrens, F.; Castellano, G. Ethnoscience, Alternative Medicine, Inflammation, and Pain Models. In *Natural Products Pharmacology and Phytochemicals for Health Care: Methods and Principles in Medicinal Chemistry*; Mahapatra, D. K.; Aguilar, C. N., Haghi, A. K., Eds.; Apple Academic–CRC: Waretown, NJ, in press.

[29] Atkins, P. *Atkins' Molecules*; Cambridge University: Cambridge, 2003.

[30] López Muñoz, F. *Book of Abstracts, Avances en el Tratamiento Farmacológico del Dolor con Analgésicos Opioides* [*Advances in the Pharmacological Treatment of Pain with Opioid Analgesics*], València, Spain, November, 5, 2019, Academy of Pharmacy of

Comunitat Valenciana–Aristo: València, Spain, Novenber, 5, 2019; RT-1.

[31] Selva Otaolaurruchi, J. *Book of Abstracts, Avances en el Tratamiento Farmacológico del Dolor con Analgésicos Opioides* [*Advances in the Pharmacological Treatment of Pain with Opioid Analgesics*], València, Spain, Novenber, 5, 2019, Academy of Pharmacy of Comunitat Valenciana–Aristo: València, Spain, Novenber, 5, 2019; RT-1.

[32] López Alarcón, M. D. *Book of Abstracts, Avances en el Tratamiento Farmacológico del Dolor con Analgésicos Opioides* [*Advances in the Pharmacological Treatment of Pain with Opioid Analgesics*], València, Spain, Novenber, 5, 2019, Academy of Pharmacy of Comunitat Valenciana–Aristo: València, Spain, 2019; RT-1.

[33] Grisolía, S.; de Andrés, J. *Book of Abstracts, El Dolor Crónico No Tiene Edad en el Siglo XXI* [*Chronic Pain Is Ageless in the 21st Century*], València, Spain, Novenber, 21, 2019, Fundación Valenciana de Estudios Avanzados: València, Spain, 2019.

[34] Borrás Blasco, C. *Book of Abstracts, El Dolor Crónico No Tiene Edad en el Siglo XXI* [*Chronic Pain Is Ageless in the 21st Century*], València, Spain, Novenber, 21, 2019, Fundación Valenciana de Estudios Avanzados: València, Spain, 2019; O-1.

[35] Mínguez Martí, A. *Book of Abstracts, El Dolor Crónico No Tiene Edad en el Siglo XXI* [*Chronic Pain Is Ageless in the 21st Century*], València, Spain, Novenber, 21, 2019, Fundación Valenciana de Estudios Avanzados: València, Spain, 2019; O-2.

[36] De Andrés, J. *Book of Abstracts, El Dolor Crónico No Tiene Edad en el Siglo XXI* [*Chronic Pain Is Ageless in the 21st Century*], València, Spain, Novenber, 21, 2019, Fundación Valenciana de Estudios Avanzados: València, Spain, 2019; O-4.

[37] Raya, Á. *Book of Abstracts, El Dolor Crónico No Tiene Edad en el Siglo XXI* [*Chronic Pain Is Ageless in the 21st Century*], València, Spain, Novenber, 21, 2019, Fundación Valenciana de Estudios Avanzados: València, Spain, 2019; O-5.

[38] Blanquer, M. *Book of Abstracts, El Dolor Crónico No Tiene Edad en el Siglo XXI* [*Chronic Pain Is Ageless in the 21st Century*], València, Spain, Novenber, 21, 2019, Fundación Valenciana de Estudios Avanzados: València, Spain, 2019; O-7.

[39] Moraleda, J.M. *Book of Abstracts*, *El Dolor Crónico No Tiene Edad en el Siglo XXI* [*Chronic Pain Is Ageless in the 21st Century*], València, Spain, Novenber, 21, 2019, Fundación Valenciana de Estudios Avanzados: València, Spain, 2019; O-6.

[40] Red de Terapia Celular TerCel. *Libro Blanco de la Terapia Celular en España* [*White Paper on Cell Therapy in Spain*]; TerCel: Madrid, Spain, 2019.

[41] Garrigues, S.; Armenta, S.; Porcar, I.; Esteve, F.; Pérez, C. *Book of Abstracts*, *Drogas en la Sociedad 4.0* [*Drugs in Society 4.0*], València, Spain, Novenber, 21-22, 2019, Universitat de València: València, Spain, 2019.

[42] Ventura, M. *Book of Abstracts*, *Drogas en la Sociedad 4.0* [*Drugs in Society 4.0*], València, Spain, Novenber, 21-22, 2019, Universitat de València: València, Spain, 2019; O-1.

[43] Giné, C. V.; Espinosa, I. F.; Vilamala, M. V. New psychoactive substances as adulterants of controlled drugs. A worrying phenomenon? *Drug Test. Anal.* 2014, 6, 819-824.

[44] Pérez, C. *Book of Abstracts*, *Drogas en la Sociedad 4.0* [*Drugs in Society 4.0*], València, Spain, Novenber, 21-22, 2019, Universitat de València: València, Spain, 2019; O-2.

[45] De San Román, U. *Book of Abstracts*, *Drogas en la Sociedad 4.0* [*Drugs in Society 4.0*], València, Spain, Novenber, 21-22, 2019, Universitat de València: València, Spain, 2019; O-3.

[46] Bravo, J. *Book of Abstracts*, *Drogas en la Sociedad 4.0* [*Drugs in Society 4.0*], València, Spain, Novenber, 21-22, 2019, Universitat de València: València, Spain, 2019; O-5.

[47] Hernández, F. *Book of Abstracts*, *Drogas en la Sociedad 4.0* [*Drugs in Society 4.0*], València, Spain, Novenber, 21-22, 2019, Universitat de València: València, Spain, 2019; O-6.

[48] Gallart, D. *Book of Abstracts*, *Drogas en la Sociedad 4.0* [*Drugs in Society 4.0*], València, Spain, Novenber, 21-22, 2019, Universitat de València: València, Spain, 2019; O-7.

[49] Costero, A. *Book of Abstracts*, *Drogas en la Sociedad 4.0* [*Drugs in Society 4.0*], València, Spain, Novenber, 21-22, 2019, Universitat de València: València, Spain, 2019; O-8.

[50] Sorribes, A. *Book of Abstracts*, *Drogas en la Sociedad 4.0* [*Drugs in Society 4.0*], València, Spain, Novenber, 21-22, 2019, Universitat de València: València, Spain, 2019; O-10.

[51] *Consumo-ConCiencia. Esto Es lo Que Hay: Leyes y Drogas; Dos Dedos de Frente* [*Consumption-Consciousness. This Is What There Is: Laws and Drugs; Two fingers at front*]: Zaragoza, Spain, 2019.

In: Innovations and Challenges in Modern … ISBN: 978-1-53618-181-4
Editor: A. K. Haghi 

*Chapter 2*

# ANTITUMOUR, PANCRACINE, PATCHOULI, ANTIFUNGAL AND CARYOPHYLLENE

***Francisco Torrens[1,*] and Gloria Castellano[2]***
[1]Institut Universitari de Ciència Molecular,
Universitat de València, València, Spain
[2]Departamento de Ciencias Experimentales y Matemáticas,
Facultad de Veterinaria y Ciencias Experimentales,
Universidad Católica de Valencia San Vicente Mártir, València, Spain

## ABSTRACT

Chinese-Herbal-Formula-CHF03 protective effect on non-alcoholic fatty liver disease in mice is related to its ability to regulate the inflammatory response and lipogenic response controlled by sterol regulatory element binding transcription factor 1. Formula CHF03 is a mix of natural compounds that could be used for the treatment of non-alcoholic fatty liver disease. Patchouli oil, alcohol, pogostone and β-patchoulene possess prebiotic-like effect in C57BL/6J mouse model. The herbal treatments improve the gut epithelial barrier, reinstating the expressions of epithelial and neural cadherin, upregulating p-*lysozyme* and *Muc 2* genes

* Corresponding Author's Email: torrens@uv.es.

expression, and suppressing the pro-inflammatory cytokine expressions. Sequencing data of 16S show that the herbal treatments positively modulate gut microbiota composition, in which the relative abundance of short-chain fatty- and lactic-acid-producing bacteria are enhanced, while certain pathogenic bacteria are reduced. Essential oils are used in a number of fields. Numerous studies highlight the action of these oils. In light of the facts that the current rise in invasive fungal infections because of the rise in immunosuppressive therapies is a real concern, and the emergence of resistant strains induces therapeutic failures, classes of antifungals are anticipated. The plant kingdom represents an immense potential of natural resources to exploit for the purposes.

**Keywords**: patchouli essential oil, patchouli alcohol, pogostone, β-patchoulene, gut microbiota, antifungal, invasive fungal infection

## 1. Introduction

Setting the scene: the bioprocess and metabolic engineering of *Cheiloclinum cognatum* roots for the production of antitumour quinonemethide triterpenes, pancracine as a novel candidate lead-structure in the therapy of oncological diseases, patchouli essential oil (EO) and its derived compounds revealing prebiotic-like effects in C57 black (C57BL)/6J mice, EOs and their natural active compounds presenting antifungal properties, caryophyllene in EO from *Chrysanthemum boreale* inducing $G_1$ phase cell cycle arrest in human lung cancer cells, natural products (NPs) in neurodegenerative diseases as a great promise but an ethical challenge, factors responsible for the risen use of NPs, a review of limonin pharmacology, toxicity and pharmacokinetics.

As an alternative therapeutic source, EOs extracted from plants are qualified as a biotechnological product to be explored for pharmaceuticals and food companies. The EOs are constituted by volatile organic products of classes synthesized by a secondary metabolism from different plants parts. Studies documented EOs to be effective antimicrobial agents *vs.* pathogens. Effect of EO is closely related to the capability of permeabilizing and/or destroying membrane integrity, leading to the leakage of intracellular substances. The antibiofilm activity of EOs was reported. Eighty percent of the world population uses traditional medicine (TM), and medicinal plants (MPs) represent a large bioactive-compounds source. Traditional Chinese medicine (TCM) compiled thousands of herbs and formulae used for disease treatment and dietary supplements. Emerging evidences showed that the certain active constituencies from TCMs possess distinct prebiotic-like effects, to which the inflammation and cancer progress were alleviated. It was speculated that TCMs might exert their medicinal functions *via* the modulation of gut microbiota (GM), and the gut epithelial environment to host benefit. β-Cell regenerative potential of selected herbal extracts was informed in alloxan-induced diabetic rats [1]. Efficacy and safety of TCM for primary intracerebral haemorrhage were reviewed [2]. Bicyclo-[6.3.0]-undecane-sesquiterpenoids structures, bioactivities and syntheses were revised [3].

Earlier publications in *Nereis*, etc. classified yams [4], lactic acid bacteria [5], fruits [6], food spices [7] and oil legumes [8] by principal component (PCA), cluster (CA) and meta-analysis (MA). The molecular classifications of 33 phenolic compounds derived from the cinnamic and benzoic acids from *Posidonia oceanica* [9], 74 flavonoids [10], 66 stilbenoids [11], 71 triterpenoids and steroids from *Ganoderma* [12], 17 isoflavonoids from *Dalbergia parviflora* [13], 31 sesquiterpene lactones (STLs) [14, 15] and STL artemisinin derivatives [16] were informed. A tool for interrogation of macromolecular structure was reported [17]. Mucoadhesive polymer hyaluronan favours transdermal penetration absorption of caffeine [18, 19]. Polyphenolic phytochemicals in cancer prevention and therapy, bioavailability and bioefficacy were reviewed [20]. From Asia to Mediterranean, soya bean, Spanish legumes and commercial *soya bean* PCA, CA and MA were published [21]. Antioxidants (AOs) of NPs from herbs and spices improved the oxidative stability and frying performance of vegetable oils [22]. The relationship between vegetable oil composition and oxidative stability was revealed *via* a multifactorial approach [23]. It was informed chemical and biological screening approaches to phytopharmaceuticals [24], cultural interbreeding in indigenous and scientific ethnopharmacology [25], ethnobotanical studies of

MPs, underutilized wild edible plants, food, medicine [26], biodiversity as a source of drugs, *Cordia*, *Echinacea*, *Tabernaemontana*, *Aloe* [27], immunomodulatory molecules from Himalayan MPs [28], antimicrobial, AO, composition of *Verbena carolina* and *Mentha* [29], phylogenesis by information entropy, avian birds and 1918 influenza virus [30]. The aim of the present work is to review bioprocess and metabolic engineering of *C. cognatum* roots for production of antitumour quinonemethide triterpenes, pancracine as a novel candidate lead-structure in therapy of oncological diseases, patchouli EO, its derived compounds revealing prebiotic-like effects in C57BL/6J mice, EOs, their natural active compounds presenting antifungal properties, caryophyllene in *C. boreale* EO inducing $G_1$ phase cell cycle arrest in human lung cancer cells, NPs in neurodegenerative diseases as a great promise but an ethical challenge, factors responsible for risen NPs use, limonin pharmacology, toxicity and pharmacokinetics.

## 2. *Cheiloclinum Cognatum* Roots for Antitumour Quinonemethide Triterpenes

Biochemical and genetic manipulation of *in vitro* root cultures are exploited to increase target-bioactives production. Roots from Celastraceae species accumulate quinonemethide triterpenoids (QTs) of interest because of their range of bioactivity, especially their notable and selective cytotoxic activity *vs.* myeloma cell proliferation, hepatocellular carcinoma, prostate cancer, glioblastoma and pancreatic cancer cells. Franca group established *in vitro* culture of *Cheiloclinum cognatum* roots, able to produce and excrete QTs, as an alternative for continuous production under monitored conditions of QTs [31]. Biochemical and genetic manipulations and strategies, e.g., exogenous addition of isotopically labelled precursors were performed to increase the yield of active metabolites, and obtain metabolic intermediates of interest for elucidation of the biosynthetic path, driving the production of *C. cognatum* QTs. Metabolomic and transcriptomic studies conducted to investigate the enzymatic regulation by oxidoreductases cytochrome P

(CYP)450, key enzymes that catalyze oxidative reactions responsible for the diversity of QTs biosynthesized in *C. cognatum* roots, allowed the identification of expressed sequences related to three CYP450 types: quinone oxidoreductase (NQO1), laccase and cytochrome C oxidoreductase. They performed analyses for sequence categorization by similarity search with sequences deposited in Genbank database (NCBI), and multiple alignments with Clustal software. *Cheiloclinum cognatum* expressed sequences related to NQO1 gene showed high similarity, more than 96%, with sequences of 12 other Celastraceae species. Enzymes of NQO1 play a role in the excretion of compounds for detoxification of cells and may be related to the excretion of QTs into *C. cognatum* root culture medium. Overall results contribute for the elucidation of key redox steps of the biosynthetic pathway of QTs, and validation of the established bioprocess as a viable alternative for the continuous production of QTs to be used in the development of innovative medicines.

## 3. Pancracine: A Novel Candidate Lead-Structure in Oncological Therapy

Cancer is one of the most dangerous worldwide diseases with lethal impacts. Thanks to a huge effort of many scientists from the whole world, the understanding of oncological diseases is considerable. Current treatment of oncological diseases includes surgical treatment, radiotherapy (RT), chemotherapy [CT, e.g., hormonal therapy (HT)], in cases immunotherapy (IT) and appropriate combination of all of them. According to National Cancer Institute (NCI), plant program is the most promising source of anticancer agents of natural origin. The NPs are a source of not only cytotoxic compounds but also substances capable of overcoming apoptosis resistance together with the anti-invasive and anti-metastatic activity. Peterova group evaluated the mechanism of anticancer action of potential therapeutics based on selected promising montanine-type isoquinoline alkaloid, pancracine, which was isolated from plant of Amaryllidaceae family at the department of Pharmaceutical Botany, Faculty of Pharmacy in Hradec Králové, and its molecule is characterized by a 5,11-methanomorphanthridine ring system [32].

## 4. Patchouli and Its Derived Compounds Revealed Prebiotic-Like Effects

*Pogostemon cablin* (PC) is a medicinal herb (MH) commonly used for treating gastrointestinal (GI) symptoms (e.g., colds, headaches, nausea, vomiting, abdominal pain, diarrhoea, ulcerative colitis, dyspepsia, poor appetite). Patchouli EO (PEO) extracted from PC leaves is the main active component and a well-known EO for a number of medicinal purposes. Patchouli alcohol (PA, $C_{15}H_{26}O$, *cf.* Figure 1a) is a tricyclic sesquiterpene [33]. It is the major component of PC and contributes to EO pungent smell. Pogostone (PO, $C_{12}H_{16}O_4$, Figure 1b) is a 2H-pyranone derived from PEO. β-Patchoulene (β-PAE, $C_{15}H_{24}$, Figure 1c), which derived from PEO, is another PEO representative hydrocarbon (HC) sesquiterpenoid.

## 5. Oils and Natural Active Compounds Presenting Antifungal Properties

The composition of an EO varies not only from one species to another, but also from one organ to another (e.g., root, aerial organs) and according to the season. It can contain tens to hundreds of different compounds. Three or four main compounds represent more than 60% of the mass and determine the bioproperties of EOs. The major components are displayed in Figure 2 [34].

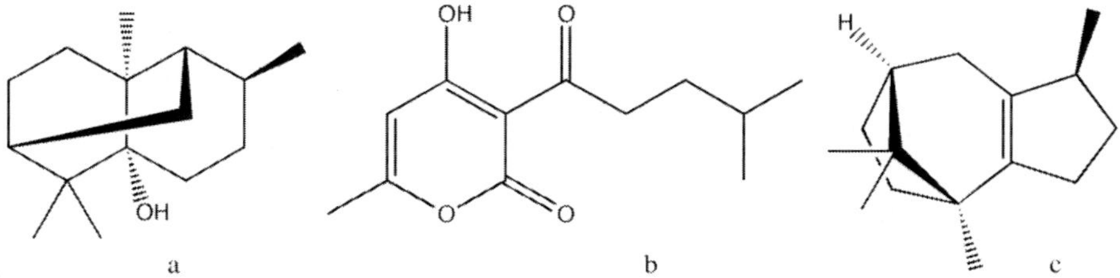


Figure 1. The chemical structure of (a) PA, (b) PO and (c) β-PAE.

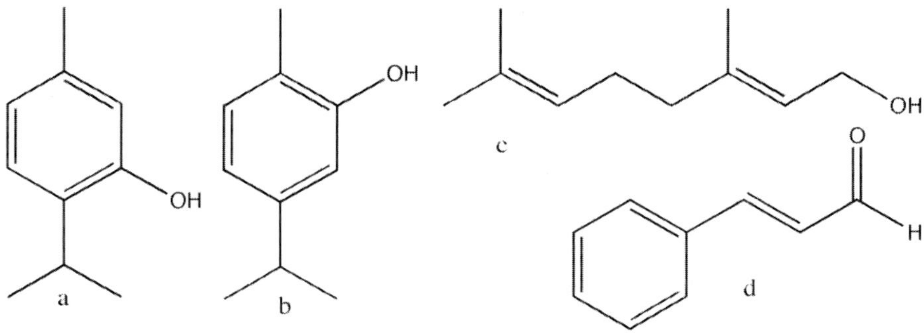


Figure 2. Major compounds of EOs: (a) thymol; (b) carvacrol; (c) geraniol; (d) (*E*)-cinnamaldehyde.

## 6. Caryophyllene in *C. Boreale* EO Induces $G_1$ Phase Cell Cycle Arrest

*Chrysanthemum boreale* (Asteraceae) is a plant widespread in East Asia, used in folk medicine to treat a number of disorders (e.g., pneumonia, colitis, stomatitis, carbuncle). Whether *C. boreale* (ECB) EO and its active constituents present anti-proliferative activities in lung cancer (LC) is unknown. The cytotoxic effects of ECB in A549 and National Cancer Institute (NCI)-H358 human LC cells were investigated [35]. Culture of A549/H358 cells with ECB induced apoptotic cell death, as revealed by a rise in annexin V staining. The ECB treatment reduced mitochondrial membrane potential (MMP), disrupted pro-anti-apoptotic Bcl-2 proteins balance and activated caspase-8/9/3, as assessed by western blot analysis. Pretreatment with a broad-spectrum caspase inhibitor (z-VAD-fmk) attenuated ECB-induced apoptosis. Gas chromatography-mass spectrometry (GC/MS) analysis of ECB identified six compounds. Among them, caryophyllene exhibited an anti-proliferative effect and was identified as the major active compound. Caryophyllene induced $G_1$ cell cycle arrest by downregulating cyclin $D_1$/E, cyclin-dependent protein kinase (CDK)-2/4/6 and retinoblastoma (RB) phosphorylation, and upregulating $p21^{CIP1/WAF1}$ and $p27^{KIP1}$, which results indicate that caryophyllene exerts cytotoxic activity in LC cells *via* induction of cell cycle arrest. Analysis of GC/MS of ECB revealed six major constituents (*cf.* Figure 3), and their cytotoxicity was determined in A549/H358 cells by 3-(4,5-dimethylthiazol-2-yl)-2,5-diphenyltetrazolium bromide (MTT) assay. Caryophyllene showed the strongest cytotoxicity among the compounds.

## 7. Degeneratives: Factors Responsible for Risen Use of Natural Products

Despite remarkable breakthroughs in people's understanding of neurodegenerative diseases (NDs), little success existed in developing effective therapies. The therapies currently available seem to be inadequate for NDs, as they only act to alleviate symptoms but cannot stop the progress of the disease. The use of NPs grows, probably because of several factors: 1) dissatisfaction with the results from traditional drugs; 2) claims on the efficacy of plant medicines; 3) rising propensity of consumers for natural therapies and alternative medicines; 4) diffuse misconception that herbal products are safer than traditional ones; 5) high cost of many traditional drugs; 6) improvements in the quality, efficacy and safety of herbal medicines with the development of science and technology; 7) risen patient decision-making autonomy regarding therapeutic options and alternatives; 8) a movement towards self-medication [36].

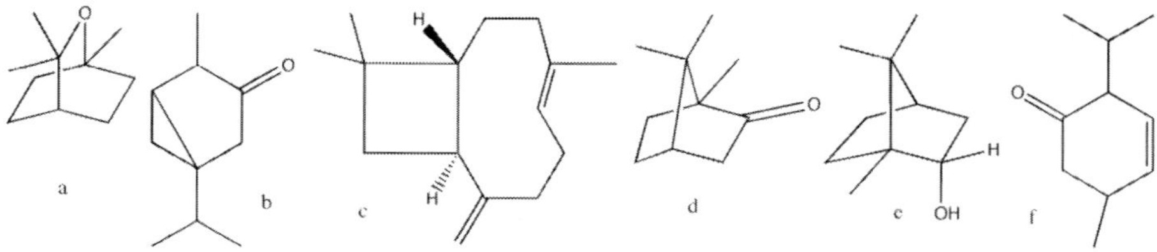


Figure 3. a) 1,8-cineole; b) thujone; c) caryophyllene; d) camphor; e) borneol; f) 2-isopropyl-5-methyl-3-cyclohexen-1-one.

## 8. Limonin: A Review of Its Pharmacology, Toxicity and Pharmacokinetics

Double δ-lactone limonin is a natural tetracyclic triterpenoid compound, which widely exists in *Euodia rutaecarpa* (Juss.) Benth., *Phellodendron chinense* Schneid. and *Coptis chinensis* Franch. Its extensive pharmacological effects attracted attention. A review provided the latest information on the pharmacology, toxicity and pharmacokinetics of limonin, exploring the therapeutic potential of the compound and looking for ways to improve efficacy and bioavailability [37]. It presents a wide spectrum of pharmacological effects [e.g., anti-cancer, anti-inflammatory (AI) and analgesic, anti-bacterial and anti-virus, AO, liver protection properties]. However, it led to hepatotoxicity, renal toxicity and genetic damage. It presents complex impacts on hepatic metabolic enzyme. Pharmacokinetic studies showed that it presents poor bioavailability, and the reduction, hydrolysis and methylation are its main metabolic pathways. The position and group of its substituents are key in affecting pharmacological activity and bioavailability. However, issues exist, e.g., its mechanism of AO activity not being clear. Few studies exist on its toxicity mechanism, and the effects of its concentration on pharmacological effects and toxicity are not clear, and no researchers reported ways in which to reduce its toxicity. It (obaculactone, evodin, *cf.* Figure 4) belongs to the tetracyclic triterpenoids and is a secondary metabolite with high bioactivity in plants.

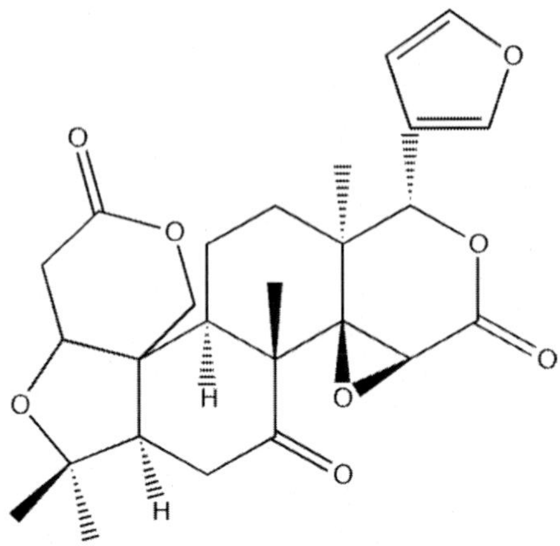


Figure 4. Molecular structure of limonin.

## 9. Discussion

The TCM alleviates chronic liver disease, reducing lipidosis, regulating immunity, improving liver function and protecting hepatocytes. *Via* 10 MHs, Chinese Herbal Formula (CHF03) was formulated to test the hypothesis that exerts protective effects *vs.* non-alcoholic fatty liver disease *in vivo* in mice [38]. Its main components were identified *via* liquid chromatography-MS analysis (amino acids, organic acids, fatty acids, nucleosides, flavones). Substances with the highest concentration were citric acid and L-arginine. Oral CFH03 reduced *in vivo/vitro* high-fat diet-induced lipid accumulation, chronic liver injury and protected hepatocytes. Patchouli is a TCM for the treatment of GI symptoms. The PEO and derived compounds are its main active ingredients that possess AI, AO, anti-microbial, anti-allergy and anti-cancer effects based on cellular and animal studies. Ten drugs are available in antifungal arsenal for the treatment of invasive fungal infections. Five classes exist, each with a different mode of action. 1) Polyenes, e.g., amphotericin B, target the plasma membrane and, in particular, ergosterol, a major component of the fungal membrane that increases membrane fluidity and causes cell death. 2) Echinocandins, e.g., caspofungin, target β-(1-3)-glucan in the fungal cell wall. 3) Azole derivatives (e.g., fluconazole, voriconazole) act on the biosynthetic pathway of ergosterol, increasing membrane fluidity and accumulation of a toxic sterol. 4) Allylamines, e.g., terbinafine, target the plasma membrane and, in particular, the first steps of ergosterol synthesis. 5) Pyrimidines, e.g., 5-fluorocytosine, act on the nucleus and, in particular, on deoxyribonucleic acid (DNA) synthesis. Mechanisms are involved in the synergistic effect of antifungals: 1) inhibition of different stages in the fungal intracellular pathways that are essential for cell survival; 2) action of another antifungal agent on the fungal cell membrane; 3) inhibition of carrier proteins; 4) simultaneous inhibition of different cell targets.

Botanical medicines were used in developing countries as the major source of treatment. Anti-cancer NPs or derivatives included 49% of 175 FDA-approved small molecules; e.g., vinca alkaloids, isolated from *Catharanthus roseus*, cause microtubule disruption and induce cell cycle arrest at metaphase, resulting in cancer-cells apoptosis. The SB365, a saponin D extracted from *Pulsatilla koreana* roots, exhibits anti-proliferative effects in cancer cell lines. In pancreatic cancer, it induces apoptosis and inhibits angiogenesis, contributing to a rise in patient survival rate to 54% without adverse effects. The efficacy of its intravenous treatment is investigated in clinical trials. *Chrysanthemum boreale* is distributed along the Korean Peninsula and spread to the Manchuria region. It and similar species (*C. indicum*, *C. lavandulaefolium*) were used in conventional Eastern treatments for stomatitis, pneumonia, carbuncle, fever and vertigo. Guaianolide derivative 8-acetoxy-4,10-dihydroxy-2,11(13)-guaiadiene-12,6-olide, isolated from *C. boreale*, presented cytotoxic activity *vs.* human cancer cell lines. However, ECB underlying mechanisms were not reported. Compounds were isolated from ECB: monoterpenes camphor, αthujone, *cis*-chrysanthenol, 1,8 cineole, α-pinene and sesquiterpene caryophyllene, which induces cell cycle arrest at $G_1$ phase in human LC cell lines, based on the observation that caryophyllene promoted downregulation of $G_1$ cell cycle positive and upregulation of $G_1$ cell cycle negative regulators.

As an alternative therapeutic source, EOs extracted from plants are qualified as an important biotechnological product to be explored for pharmaceuticals and food companies [39]. The EOs are constituted by volatile organic products of classes (e.g., aldehydes, terpenes, phenolic compounds) synthesized by a secondary metabolism from different plants parts. Studies documented EOs to be effective antimicrobial agents *vs.* pathogens. Effect of EO is closely related to the capability of permeabilizing and/or destroying membrane integrity, leading to the leakage of intracellular substances. Antibiofilm activity of EOs was published. The growing threat of antimicrobial drug resistance to conventional antimicrobial drugs stimulated the search for therapeutic alternatives (e.g., extracts of MPs, EOs), which are recognized for potential antimicrobial role *vs.* microorganisms [40]. Eighty percent of the world population uses TM, and MPs represent a large bioactive-compounds source. Frankincense and myrrha represent a clinically typical herbal pairing found in *Ruxiang Zhitong San*, with effects of invigorating circulation, reducing stasis, astringing wounds and generating muscle [41].

## CONCLUSION

From the present results and discussion, the following final remarks can be drawn.

1. Chinese-Herbal-Formula-CHF03 protective effect on non-alcoholic fatty liver disease in mice is related to its ability to regulate the inflammatory response and lipogenic response controlled by sterol regulatory element binding transcription factor 1. Thus, Formula CHF03 is a promising mix of natural compounds that could be used for the treatment of non-alcoholic fatty liver disease.
2. Patchouli oil/alcohol/pogostone/β-patchoulene possess prebiotic-like effect in C57BL/6J mouse model. The herbal treatments improved the gut epithelial barrier, reinstating the expressions of epithelial and neural cadherin, upregulating p-*lysozyme* and *Muc 2* genes expression, and suppressing the pro-inflammatory cytokine expressions. Sequencing data of 16S showed that the herbal treatments positively modulated gut microbiota composition, in

which the relative abundance of short-chain fatty- and lactic-acid-producing bacteria were enhanced, while certain pathogenic bacteria were reduced.

3. Essential oils are widely used in a number of fields, e.g., cosmetology or the food industry. Numerous studies highlighted the action of these oils on various physiological effects in humans. In addition, their antimicrobial effects were recognized.
4. The antifungal effects of these oils, alone or in combination with pre-existing antifungals, would be an interesting solution for improving therapeutic failures related to emergence of resistant strains and an increase in immunosuppressive treatments in medicine.
5. Although the studies are promising, people must not forget to consider the toxicity of these products. In fact, essential oils are already known to pose a high risk of allergy and irritation.
6. Oil from *Chrysanthemum boreale* exerts its anti-proliferative effect *via* caspase activation and mitochondria-dependent apoptosis in human lung cancer cells. Its active constituent is caryophyllene, which promotes $G_1$ cell cycle arrest in the cell types. The oil is a potential chemotherapeutic agent for the treatment of non-small cell lung cancer.

## Acknowledgments

The authors thank support from Fundación Universidad Católica de Valencia *San Vicente Mártir* (Project No. 2019-217-001UCV).

## References

[1] Attanayake, A. P.; Jayatilaka, K. A. P. W.; Mudduwa, L. K. B.; Pathirana, C. β-cell regenerative potential of selected herbal extracts

in alloxan induced diabetic rats. *Curr. Drug Discov. Technol.* 2019, 16, 278-284.

[2] Wang, H. L.; Zeng, H.; Xu, M. B.; Zhou, X. L.; Rong, P. Q.; Jin, T. Y.; Wang, Q.; Zheng, G. Q. Efficacy and safety of Chinese herbal medicine for primary intracerebral hemorrhage: A systematic review of randomized controlled trials. *Front. Pharmacol.* 2019, 10, 1139-1-14.

[3] Qin, G. F.; Liang, H. B.; Liu, W. X.; Zhu, F.; Li, P. L.; Li, G. Q.; Yao, J. C. Bicyclo [6.3.0] undecane sesquiterpenoids: Structures, biological activities, and syntheses. *Molecules* 2019, 24, 3912-1-11.

[4] Torrens-Zaragozá, F. Molecular categorization of yams by principal component and cluster analyses. *Nereis* 2013, 2013(5), 41-51.

[5] Torrens-Zaragozá, F. Classification of lactic acid bacteria against cytokine immune modulation. *Nereis* 2014, 2014(6), 27-37.

[6] Torrens-Zaragozá, F. Classification of fruits proximate and mineral content: Principal component, cluster, meta-analyses. *Nereis* 2015, 2015(7), 39-50.

[7] Torrens-Zaragozá, F. Classification of food spices by proximate content: Principal component, cluster, meta-analyses, *Nereis* 2016, 2016(8), 23-33.

[8] Torrens, F.; Castellano, G. From Asia to Mediterranean: Soya bean, Spanish legumes and commercial *soya bean* principal component, cluster and meta-analyses. *J. Nutr. Food Sci.* 2014, 4(5), 98-98.

[9] Castellano, G.; Tena, J.; Torrens, F. Classification of polyphenolic compounds by chemical structural indicators and its relation to antioxidant properties of *Posidonia oceanica* (L.) Delile. *MATCH Commun. Math. Comput. Chem.* 2012, 67, 231-250.

[10] Castellano, G.; González-Santander, J. L.; Lara, A.; Torrens, F. Classification of flavonoid compounds by using entropy of information theory. *Phytochemistry* 2013, 93, 182-191.

[11] Castellano, G.; Lara, A.; Torrens, F. Classification of stilbenoid compounds by entropy of artificial intelligence. *Phytochemistry* 2014, 97, 62-69.

[12] Castellano, G.; Torrens, F. Information entropy-based classification of triterpenoids and steroids from *Ganoderma. Phytochemistry* 2015, 116, 305-313.

[13] Castellano, G.; Torrens, F. Quantitative structure-antioxidant activity models of isoflavonoids: A theoretical study. *Int. J. Mol. Sci*. 2015, 16, 12891-12906.

[14] Castellano, G.; Redondo, L.; Torrens, F. QSAR of natural sesquiterpene lactones as inhibitors of Myb-dependent gene expression. *Curr. Top. Med. Chem*. 2017, 17, 3256-3268.

[15] Torrens, F.; Castellano, G. Structure-activity relationships of cytotoxic lactones as inhibitors and mechanisms of action. *Curr. Drug Discov. Technol*., in press.

[16] Torrens, F.; Redondo, L.; Castellano, G. Artemisinin: Tentative mechanism of action and resistance. *Pharmaceuticals* 2017, 10, 20-4-4.

[17] Torrens, F.; Castellano, G. A tool for interrogation of macromolecular structure. *J. Mater. Sci. Eng. B* 2014, 4(2), 55-63.

[18] Torrens, F.; Castellano, G. Mucoadhesive polymer hyaluronan as biodegradable cationic/zwitterionic-drug delivery vehicle. *ADMET DMPK* 2014, 2, 235-247.

[19] Torrens, F.; Castellano, G. Computational study of nanosized drug delivery from *cyclo*dextrins, crown ethers and hyaluronan in pharmaceutical formulations. *Curr. Top. Med. Chem*. 2015, 15,1901-1913.

[20] Estrela, J. M.; Mena, S.; Obrador, E.; Benlloch, M.; Castellano, G.; Salvador, R.; Dellinger, R. W. Polyphenolic phytochemicals in cancer prevention and therapy: Bioavailability *versus* bioefficacy. *J. Med. Chem*. 2017, 60, 9413-9436.

[21] Torrens, F.; Castellano, G. From Asia to Mediterranean: Soya bean, Spanish legumes and commercial *soya bean* principal component, cluster and meta-analyses. *J. Nutr. Food Sci*. 2014, 4(5), 98-98.

[22] Redondo-Cuevas, L.; Castellano, G.; Raikos, V. Natural antioxidants from herbs and spices improve the oxidative stability and frying

performance of vegetable oils. *Int. J. Food Sci. Technol.* 2017, 52, 2422-2428.

[23] Redondo-Cuevas, L.; Castellano, G.; Torrens, F.; Raikos, V. Revealing the relationship between vegetable oil composition and oxidative stability: A multifactorial approach. *J. Food Compos. Anal.* 2018, 66, 221-229.

[24] Torrens, F.; Castellano, G. Chemical/Biological Screening Approaches to Phytopharmaceuticals. In: *Research Methods and Applications in Chemical and Biological Engineering*; Pourhashemi, A., Deka, S. C., Haghi, A. K., Eds.; Apple Academic-CRC: Waretown, NJ, in press.

[25] Torrens, F.; Castellano, G. Cultural Interbreeding in Indigenous/Scientific Ethnopharmacology. In: *Research Methods and Applications in Chemical and Biological Engineering*; Pourhashemi, A., Deka, S. C., Haghi, A. K., Eds.; Apple Academic-CRC: Waretown, NJ, in press.

[26] Torrens, F.; Castellano, G. Ethnobotanical Studies of Medicinal Plants: Underutilized Wild Edible Plants, Food and Medicine. In: *Innovations in Physical Chemistry*; Haghi, A. K., Ed.; Apple Academic-CRC: Waretown, NJ, in press.

[27] Torrens, F.; Castellano, G. Biodiversity as a source of drugs: Cordia, Echinacea, Tabernaemontana *and* Aloe. In: *Green Chemistry and Biodiversity: Principles, Techniques, and Correlations*; Aguilar, C. N., Ameta, S. C., Haghi, A. K., Eds.; Apple Academic-CRC: Waretown, NJ, in press.

[28] Torrens, F.; Castellano, G. Immunomodulatory Molecules from Himalayan Medicinal Plants. In: *Chemistry and Chemical Engineering for Sustainable Development: Best Practices and Research Directions*; Esteso, M. A., Ribeiro. A. C. F., Haghi, A. K., Eds.; Apple Academic-CRC: Waretown, NJ, in press.

[29] Torrens, F.; Castellano, G. Antimicrobial, Antioxidant, and Composition of Verbena carolina and Mentha. In: *Applied Pharmaceutical Science and Microbiology: Novel Green Chemistry*

*Methods and Natural Products*; Mahapatra, D. K., Talele, S. G., Haghi, A. K., Eds.; Apple Academic-CRC: Waretown, NJ, in press.

[30] Torrens, F.; Castellano, G. Phylogenesis by Information Entropy: Avian Birds and 1918 Influenza Virus. In: *Modelling & Simulation*; Turner, S., Yunus, J., Eds.; IEEE: London, 2008; Vol. 2: pp 1-2.

[31] Franca, S. C.; Pina, E. S.; Pereira, A. M. S.; Lopes, A. A.; Coppede, J. S., *Book of Abstracts, World Congress on Biotechnology and Biological Studies*, València, Spain, October 24-25, 2019, Conference Era: València, Spain, 2019; TS-1.

[32] Peterova, E.; Koutova, D.; Havelek, R.; Breiterova, K.; Cahlíková, L.; Rezacova, M. *Book of Abstracts, World Congress on Biotechnology and Biological Studies*, València, Spain, October 24-25, 2019, Conference Era: València, Spain, 2019; P-1.

[33] Leong, W.; Huang, G.; Khan, I.; Xia, W.; Li, Y.; Liu, Y.; Li, X.; Han, R.; Su, Z.; Hsiao, W. L. W. Patchouli essential oil and its derived compounds revealed prebiotic-like effects in C57BL/6J mice. *Front. Pharmacol.* 2019, 10, 1229-1-11.

[34] D'agostino, M.; Tesse, N.; Frippiat, J. P.; Machouart, M.; Debourgogne, A. Essential oils and their natural active compounds presenting antifungal properties. *Molecules* 2019, 24, 3713-1-22.

[35] Chung, K. S.; Hong, J. Y.; Lee, J. H.; Lee, H. J.; Park, J. Y.; Choi, J. H.; Park, H. J.; Hong, J.; Lee, K. T. β-Caryophyllene in the essential oil from *Chrysanthemum boreale* induces $G_1$ phase cell cycle arrest in human lung cancer cells, *Molecules* 2019, 24, 3754-1-11.

[36] Di Paolo, M.; Papi, L.; Gori, F.; Turillazzi, E. Natural products in neurodegenerative diseases: A great promise but an ethical challenge. *Int. J. Mol. Sci.* 2019, 20, 5170-1-12.

[37] Fan, S.; Zhang, C.; Luo, T.; Wang, J.; Tang, Y.; Chen, Z.; Yu, L. Limonin: A review of its pharmacology, toxicity, and pharmacokinetics, *Molecules* 2019, 24, 3679-1-22.

[38] Cui, Y.; Chang, R.; Zhang, T.; Zhou, X.; Wang, Q.; Gao, H.; Hou, L.; Loor, J. L.; Xu, C. Chinese Herbal Formula (CHF03) attenuates non-alcoholic fatty liver disease (NAFLD) through inhibiting

lipogenesis and anti-oxidation mechanisms. *Front. Pharmacol.* 2019, 10, 1190-1-12.

[39] Melo, R. S.; Azevedo, Á. M. A.; Pereira, A. M. G.; Rocha, R. R.; Cavalcante, R. M. B.; Matos, M. N. C.; Lopes, P. H. R.; Gomes, G. A.; Rodrigues, T. H. S.; dos Santos, H. S.; Ponte, I. L.; Costa, R. A.; Brito, G. S.; Catunda, F. E. A., Jr.; Carneiro, V. A. Chemical composition and antimicrobial effectiveness of *Ocimum gratissimum* L. essential oil against multidrug-resistant isolates of *Staphylococcus aureus* and *Escherichia coli*. *Molecules* 2019, 24, 3864-1-17.

[40] Tullio, V.; Roana, J.; Scalas, D.; Mandras, N. Enhanced killing of *Candida krusei* by polymorphonuclear leucocytes in the presence of subinhibitory concentrations of *Melaleuca alternifolia* and *Mentha of Pancalieri* essential oils. *Molecules* 2019, 24, 3824-1-10.

[41] Miao, X. D.; Zheng, L. J.; Zhao, Z. Z.; Su, S. L.; Zhu, Y.; Guo, J. M.; Shang, E. X.; Qian, D. W.; Duan, J. A. Protective effect and mechanism of boswellic acid and myrrha sesquiterpenes with different proportions of compatibility on neuroinflammation by LPS-induced BV2 cells combined with network pharmacology. *Molecules* 2019, 24, 3946-1-16.

In: Innovations and Challenges in Modern … ISBN: 978-1-53618-181-4
Editor: A. K. Haghi 

*Chapter 3*

# REVIEW: POSSIBLE FIELDS OF METAL CARBON MESOSCOPIC COMPOSITES APPLICATION

***V. I. Kodolov*[1,2,*]*, V. V. Kodolova-Chukhontseva*[1,3]*, I. N. Shabanova*[1,3]*, N. S. Terebova*[1,3]*, Yu. V. Pershin*[1,2]*, R. V. Mustakimov*[1,2]* and A. Yu. Bondar*[2]**

[1]Basic Research – High Educational Centre of Chemical Physics and Mesoscopics, UD, RAS, Izhevsk, Russia
[2]M. T. Kalashnikov Izhevsk State University, Izhevsk, Russia
[3]Udmurt Federal Research Centre, Izhevsk, Russia

## ABSTRACT

This chapter is dedicated to the consideration of the metal carbon mesocomposites application possibilities for the different practical trends. Each from these trends is determined by correspondent peculiarities of content and structure of mesoscopic composites. The main peculiarities of these nanosized particles are following: a) the presence of unpaired

* Corresponding Author's Email: vkodol.av@mail.ru.

electrons on the carbon shell; b) the structure of carbon shell from poly acetylene and carbine fragments; c) the atomic magnetic moment of inner metal is equaled to more than 1–3 $\mu_B$. The creation of reactive mesoscopic materials with regulated magnetic characteristics which can find the application as modifiers of materials properties, catalysts for different processes, effective inhibitors of oxidation processes including corrosion processes, sorbents, stimulators of plant growth and also as medicine magnetic transport within organisms, is very topical. The present investigation has fundamental character. It's based on the ideas concerning to the change of metal carbon mesocomposites reactivity. The using is possible as metal carbon mesocomposites both and their modified analogous. The application examples are presented for the properties improvement of concretes, polymer compounds, plastics as well as for the increasing of positive qualities of such substances as glues, corrosion inhibitors, fire retardants, stimulators of plants growth and medicine magnetic transport.

**Keywords**: metal carbon mesoscopic composites (mesocomposites), phase coherency, charge quantization, interference, annihilation, self organization, red ox synthesis, catalysts, reagents, modifiers, adhesives, glues, inorganic and organic polymeric materials, corrosion inhibitors, fire retardants, magnetic transport of medicines in organism, stimulators of plants growth

## 1. INTRODUCTION

The manuscript is presented as the review of series of papers, manuscripts and patents on the obtaining, investigations and applications of uncials magnetic mesoparticles which are mesoscopic metal carbon composites [1–41]. The correspondent structure and content of these mesocomposites are caused by the special conditions of their production. In as much as the application of mesoparticles is determined by the properties of these particles which are defined by their structure and content. At the beginning let us consider the initial metal carbon mesocomposite electron structure and correspondent magnetic characteristics (metal atomic magnetic moments, and the spin quantities on carbon shell). The activity of metal-carbon mesoscopic composites are caused by the structure and composition of correspond mesocomposites, which contain the delocalized electrons and double bonds on the surface of carbon shell. Therefore the initial mesoscopic composites are early participated in reactions, especially radial processes and reduction oxidation processes. This activity may be used in modification processes mesocomposites accompanied by the magnetic characteristics changes in modified mesoscopic composites. That development of possibilities processes opens new era for further investigations and development of metal carbon mesoscopic composites application fields. The expansion of metal carbon mesocomposites application chances takes place.

## 2. Compositions and Structures of Metal Carbon Mesocomposites Including Modified Mesocomposites

The representation about metal carbon mesocomposites is based on the results of the following investigations: IR or Raman spectroscopy, x-ray photoelectron spectroscopy (XRPES), electron paramagnetic resonance (EPR), transition electron microscopy (TEM), atomic force microscopy (AFM). The images of Copper Carbon and Nickel Carbon mesocomposites structures show on Figure 1a, b [33, 41].

According to the investigation results the Copper Carbon mesocomposite includes Copper containing clusters linked with Carbon shell which consists from 3-4 layers of carbon fibers. These fibers contain the carbine and poly acetylene fragments on the joints which the unpaired electrons are found. These fragments are defined by means of x-ray photoelectron spectra (Figure 2a, b) as well as metals atomic magnetic moments.

The atomic magnetic moment of Copper is equaled to 1,3 $\mu_B$. The mesocomposite obtained is tested as materials modifier (in minute quantities, less than 0,01%), as additives to corrosion inhibitors, fire retardants, thermal retardants, antistatic as well as catalysts or reagents at radical or reduction–oxidation processes, stimulators of plant growth, magnetic transport of medicine in organism. Further the examples of Copper Carbon mesocomposite application are presented.

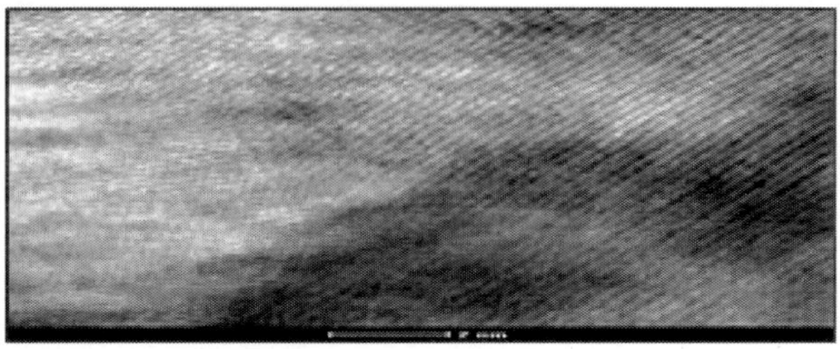

a

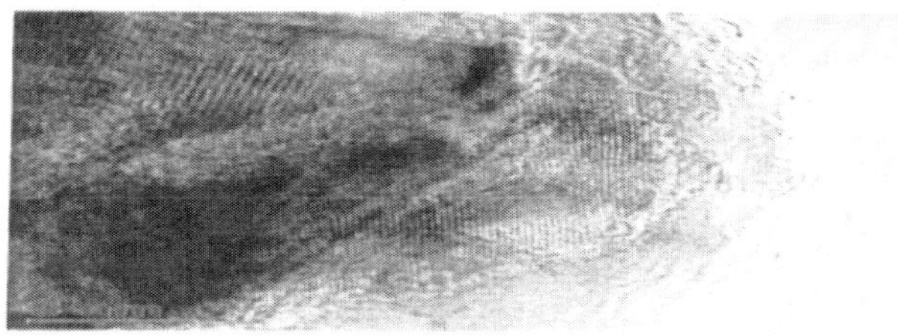

b

Figure 1. TEM microphotographs for Copper Carbon (a) and Nickel Carbon (b) mesocomposites.

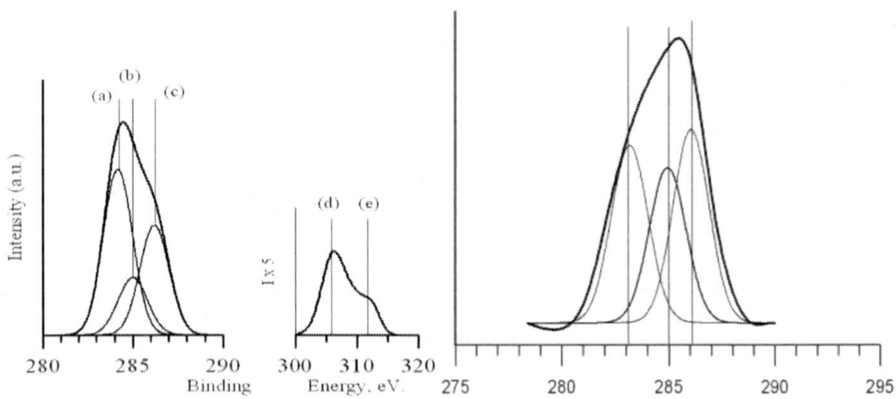


a b

Figure 2. C1s spectra for Copper Carbon (a) and Nickel Carbon mesocomposites.

Nickel Carbon mesocomposite has similar structure to previous Copper Carbon mesoscopic composite only with the increased quantity of joints between carbine and poly acetylene fragments and consequently more number of unpaired electrons. This mesocomposite may be ascribed to ferromagnetic substances. Atomic magnetic moment of mesocomposite is equaled to 1,8 $\mu_B$.

According to the XPS spectra (Figure 2), the carbon fibers can contain fragments of carbine and poly acetylene; the component at 281-282 eV is ascribed to carbine, and the peak at 285 eV is attributed to the C-H bond. It can be assumed that at the junction of the two fragments the formation of unpaired electrons is possible. The coordination interaction of metal with double bonds of corresponding fragments, mainly carbine fragments, can be a stabilizing factor for such carbon shell. From the C1s spectra for the Cu-C and Ni-C mesocomposites it can be seen that the peak corresponding to C-H bonds is 30% higher for the Ni-C mesocomposite than that for the Cu-C mesocomposite. This fact can be connected with the decrease (shortening) of carbine fragments in the Ni-C mesocomposite. The obtained result can be explained by the incomplete reduction of nickel oxide during the Ni-C mesocomposite formation (Table 1).

Table 1 presents the ratio of metal-containing phases in the metal carbon mesocomposite calculated with the use of the X-ray diffraction patterns [31].

**Table 1. The metal containing phases content in Cu-C and Ni-C mesocomposites**

| Phase | Cu-C mesocomposite | Ni-C mesocomposite |
|---|---|---|
| CuO | 1.17% | - |
| $Cu_2O$ | 5.19% | - |
| Cu | 93.64% | - |
| NiO | - | 32.15% |
| Ni | - | 67.85% |

From the previous it is following that the Ni-C mesocomposite has more junctions between the carbine and poly acetylene fragments and, consequently, a larger number of unpaired electrons that is confirmed by the

EPR spectra: for the Ni-C mesocomposite – $2.46 \cdot 10^{23}$ spin/g, and for the Cu/C mesocomposite – $1.2 \cdot 10^{17}$. The presence of unpaired electrons on the fibers of carbon shell of the studied particles indicates the possibility of their activity in the reactions with chemical substances. The results of TEM and x-ray photoelectron spectroscopy investigations are very well confirmed to EPR spectra (Table 2). EPR spectrum of Copper Carbon mesocomposite is presented as a singlet spectrum in which the distance between points of incline maximum $\Delta H_{pp}$ = 6,8 Hz, g-factor equals to g-factor of biphenyl picryl hydrazyl (g = 2,0036), and the unpaired electrons number corresponds to value $1{,}2 \cdot 10^{17}$ spin/g. In the comparison with this spectrum, the EPR spectrum for Nickel/Carbon mesocomposite has $\Delta H_{pp}$ corresponding to 2400 Hz, g – 2,46 and the unpaired electrons number – $10^{22}$ spin/g.

**Table 2. Data of EPR spectra for Copper Carbon and Nickel Carbon Mesocomposites**

| Type of metal carbon mesocomposite | g-factor | Number of unpaired electrons, spin/g |
|---|---|---|
| Copper Carbon mesomposite | 2,0036 | $1{,}2 \cdot 10^{17}$ spin/g |
| Nickel Carbon mesocomposite | 2,46 | $10^{23}$ spin/g |

The difference in EPR spectra is in all probability explained by the Carbon shell difference of these mesocomposites.

Cobalt Carbon mesoscopic composite is obtained with following characteristics: the density – 1,9 g/cm$^3$, the Cobalt atomic magnetic moment is equaled to 1,6 $\mu_B$, and also this mesocomposite has the high chemical activity at concentration less than 0,01% in polar media.

Iron Carbon mesocomposite has the following characteristics: the density – 2,1 g/cm$^3$, the Iron atomic magnetic moment – 2,5 $\mu_B$, the high chemical activity in polar media at concentration equaled to 0,001%.

Both these composites have analogous structures similar to structures of previous Ni-C and Cu-C mesocomposites. These substances may be classified as stable radicals with the migration of unpaired electrons from metal to carbon shell and back. These mesoparticles can stimulate the transport of electrons within media. The correspondent structures and properties of mesocomposites cause their application as catalysts and active reagents or additives in the different reactions.

## 3. CATALYSTS AND REAGENTS FOR DIFFERENT PROCESSES

### 3.1. Reagents for Reduction-Oxidation Processes

The investigation of Copper- and Nickel Carbon mesocomposites (Cu-C and Ni-C MC) interactions with such reagents as ammonium polyphosphate (APPh), silica ($SiO_2$), aluminium oxide, iron oxide, nickel oxide and copper oxide is carried out. In correspondent reactions the element reduction for reagents and Nickel or Copper atomic magnetic moments growth in mesoparticles take place. Below in Tables 3 and 4 the examples of metal atomic magnetic moments changes (in Boron magnetons) and quantities of unpaired electrons (in spin/g) for mesoparticles modified by APPh or silica after the mechanochemical modification processes proceeding are given.

The metal atomic magnetic moment growth proceeds owing to the redox processes with above chemical compounds. In papers [32, 33] it's shown that the reduction reactions of Phosphorus and Silicon from correspondent substances at the interaction on the interphase boundary with mesoparticles are realized (Figure 3, 5).

**Table 3. The values of Copper (Nickel) atomic magnetic moments in the interaction products for systems: Cu-C MC – APPh (or $SiO_2$) and Ni-C MC – APPh (or $SiO_2$)**

| Systems Cu-C MC – substances | $\mu_{Cu}$ | Systems Ni-C MC - substance | $\mu_{Ni}$ |
|---|---|---|---|
| Cu-C MC – silica | 3,0 | Ni-C MC – silica | 4,0 |
| Cu-C MC – APPh | 2,0 | Ni-C MC – APPh | 3,0 |
| Cu-C MC - APPh, relation 1:0,5 | 4,2 | | |

**Table 4. The unpaired electron values (from EPR spectra) for systems "Cu-C MC – silica" and "Cu-C MC – APPh" (relation 1:1) in comparison with initial mesoparticle Cu-C MC**

| Substance | Quantity of unpaired electrons, spin/g |
|---|---|
| Cu-C mesocomposite | $1{,}2 \times 10^{17}$ |
| system «Cu-C MC – $SiO_2$» | $3{,}4\times10^{19}$ |
| system «Cu-C MC – APPh» | $2{,}8\times10^{18}$ |

The increase of the metal atomic magnetic moment during the mixing of the Cu-C mesocomposite with chemical compounds and the change of the number of delocalized electrons on carbon structures is connected with the interaction of a nanosized particle with the environment. The assumed mechanism of the increase of the metal atomic magnetic moment is the electron exchange at the phase boundary similar to heterogeneous catalysis in multi cells. For a large particle, the relation of the surface to mass is so small that the metal surrounding cannot influence the electron density of its d-zone; in this case, the magnetic moment of a small particle can noticeably change. It is shown in [33] that as a result of mechanochemical mixing, in the obtained systems "Cu-C – ammonium polyphosphate" and "Cu-C – silica," phosphorus or silicon of the compound are reduced; this fact confirms the interaction presence of mesocomposite with reagents on the phase boundary.

The XPS P2p spectra for phosphorus-containing mesocomposites prepared at the ratio of presents the reagents Cu-C: ammonium polyphosphate = 1 and Cu-C: ammonium polyphosphate = 2 show in Figure 3.

The phosphorus reduction at the ratio of 2 is almost 50% more complete [32, 33] and from the comparison Figure 3a and Figure 3b.

Such result can be explained by the rate of the electron quantization at the increase in the thickness of the ammonium polyphosphate layer.

In this case the structure Phosphorus containing Copper Carbon mesocomposite is similar to the initial mesocomposite (Figure 4), only the increasing of carbon fibers thickness is noted because of the possible disposition of phosphorus reduced between carbon atoms of carbon fibers.

The Phosphorus disposition between carbon fibers can be explained by the interaction of Phosphorus electrons with the carbine double bonds electrons that is confirmed by further studies on the reactions of mesocomposite obtained with different reagents.

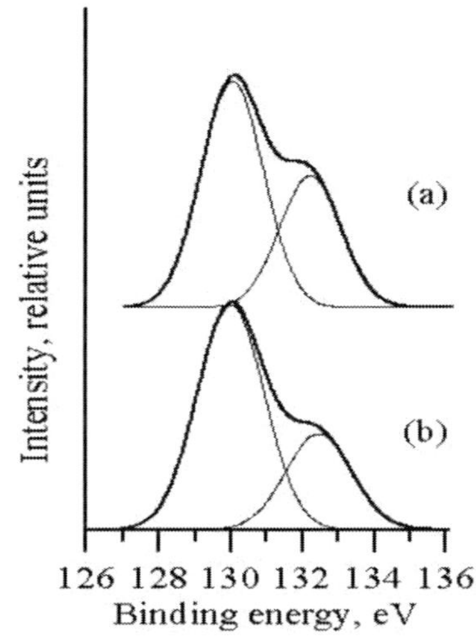


Figure 3. The XPS P2p spectra: a) Cu/C + APP 1:1; b) Cu/C + APP 1:0.5 [8].

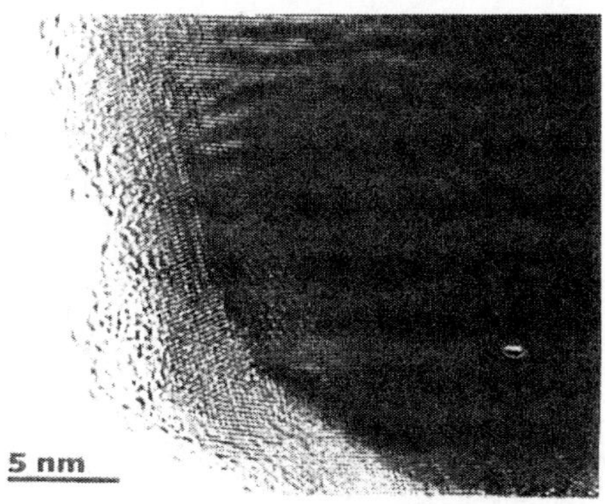


Figure 4. TEM microphotograph for Phosphorus containing Copper Carbon mesoocomposite.

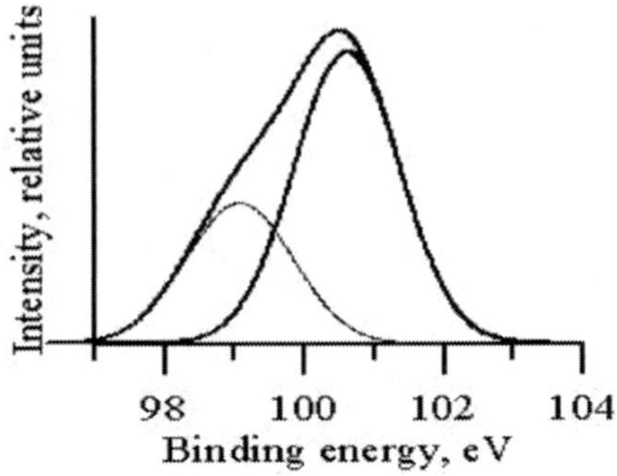


Figure 5. The XPS Si2p spectrum for the sample Cu/C + $SiO_2$ at the ratio 1:1 (E(SiO) = 100.7 eV; E(Si) = 99 eV).

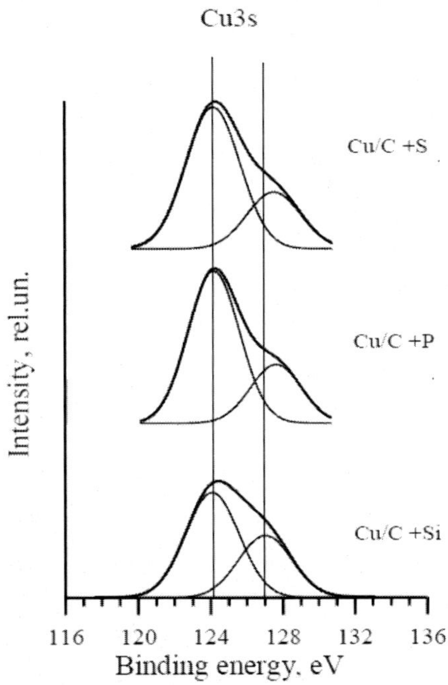


Figure 6. x Ray photoelectron spectra of Copper-Carbon mesocomposites modified by p element containing substances.

In the case, when ammonium polyphosphate is replaced by aerosol (or silicon), the reduction of silicon from silicon oxide takes place by 50% independent of the used amount of the reagents (Figure 5); this can be explained by the decrease of the rate of the electron quantization in the layer containing Si-O bonds.

The early created method [2, 7] on the base of the x-ray photoelectron spectra is used for the Copper Carbon mesocomposite modification mechanism investigation. The inner levels spectra for C1s, O1s, Si2p, P2p, S2p Cu3s are studied to investigate the mechanism of chemical bonds formation between metal or carbon atoms and silicon or phosphorus atoms. It's noted the presence of changes in the Cu3s spectra in mesocomposites modified in comparison with initial mesocomposite.

Cu3s spectra of mesoparticles modified by Si, P and S containing compounds are presented below (Figure 6).

Below the results of investigations by means of x-ray electron spectscopy for Nickel Carbon nanocomposite interactions with oxides of such metals as Aluminium, Iron, Nickel and Copper are given in Tables 5-7 and in Figures 7, 8.

The relations of nanocomposites (mesoparticles) to above oxides are changed from 1:1 to 1:0,2 depending on the qualitive spectra obtaining, For example, the relations of 1:1 and 1:0,5 for system "Ni-C MC – $Al_2O_3$" leads to full mask of mesoparticle. Therefore the quantity of aluminium oxide is decreased to the relation 1 : 0,2. In accordence with Al3s spectra Aluminum is completely reduced during the modification process, and Nickel atomic magnetic moment is increased to 4,8$\mu_B$ (Table 5).

**Table 5. Parameters of multiple splintering Me3s spectra in system Ni-C + $Al_2O_3$**

| Sample | $I_2/I_1$ (Ni) | $\Delta_{Ni}$, eV | $\mu_{Ni}$ $\mu_Б$ |
|---|---|---|---|
| Ni/C | 0.3 | 3.0 | 1.8 |
| Ni/C + $Al_2O_3$=1:0.2 | 0.9 | 1.0 | 4.8 |

$I_2/I_1$ – the relation of multiple splintering lines maximums intensities;

Δ - Energetic distance between the multiple splintering maximums in Me3s spectra.

**Table 6. Parameters of multiple splintering Me3s spectra in systems "Ni-C MC – $Fe_2O_3$" and "Ni-C MC – CuO"**

| Sample | $I_2/I_1$ (Ni) | $\Delta_{Ni}$, eV | $\mu_{Ni}$ $\mu_Б$ | $(I_2/I_1)_{Fe}$ | $\Delta_{Fe}$, eV | $\mu_{Fe}$,$\mu_Б$ |
|---|---|---|---|---|---|---|
| Ni-C MC + ($Fe_2O_3$) | 0.5 | 2.4 | 2.5 | 0.6 | 2.0 | 3.0 |
| Ni-C MC + CuO | 0.5 | 3.0 | 2.5 | 0.4 | 3.0 | 2.0 |

$I_2/I_1$ – the relation of multiple splintering lines maximums intensities;
Δ - Energetic distance between the multiple splintering maximums in Me3s spectra.

In this case the reduction process is related to not only Aluminum oxide but also to Nickel oxide from metal cluster of mesoparticle (Ni-C MC). Therefore the reduction processes are stipulated by the electron transport from carbon shell of mesoparticle in direction to $Al^{+3}$ and $Ni^{+2}$ of atoms in correspondent oxides.

The next researches with using of x-ray photoelectron spectroscopy are carried out for the following systems: Ni-C MC – $Fe_2O_3$ and Ni-C MC – CuO (Table 6).

It's necessary to note, that the iron oxide is sufficient easily transformed in $Fe_3O_4$ (magnetite). At the heating the farther reduction of iron with its implantation within the carbon shell is observed.

When Copper oxide (CuO) is added to Ni-C mesocomposite in the relation 1:1 at the grinding, the relative intensity of multiple maximum for Ni3s and Cu3s spectra is correlated with the atomic magnetic moments values (Table 6).

C1s spectrum for system "Ni-C MC–CuO" consists from 3 components: C–C ($sp^2$ hybridization) 284 eV; C–H bond 285 eV; C–C ($sp^3$ hybridization) 286 eV. The results of multiple splintering in Ni3s spectra for systems "Ni-C MC –CuO" and "Ni-C MC – NiO" are closed (Tables 6, 7).

X ray photoelectron spectroscopy investigations of Ni/C mesoparticles modified by Nickel oxide are carried out for the following systems: a) Ni-C MC – NiO, relation 1:1; b) Ni-C MC – NiO, relation 1:0,5. The researches proceed without the sample heating and with the heating (300°C) of samples. When the studies are realized without heating the intensity of maximum Ni2p line corresponds to binding energy equaled to 855 eV (NiO). However at the heating to 300ºC the intensity maximum for this line is attributed on binding energy to 852 eV (Ni). Below the Ni3s spectra for a) and b) systems without heating are shown (Figure 7).

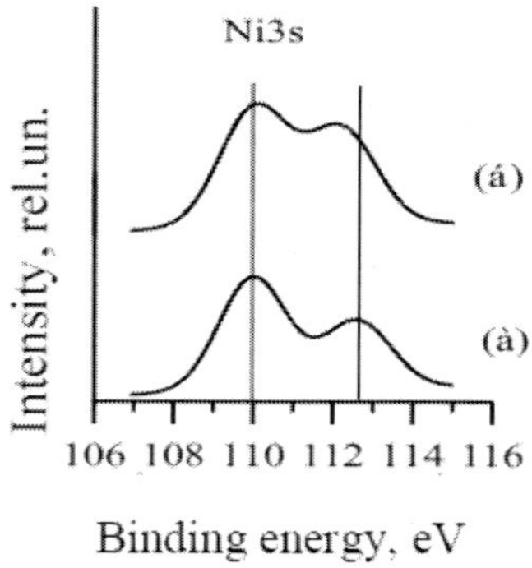


Figure 7. Ni3s spectra for systems: a) Ni C MC – NiO (1:0,5);
b) Ni-C MC – NiO (1:1).

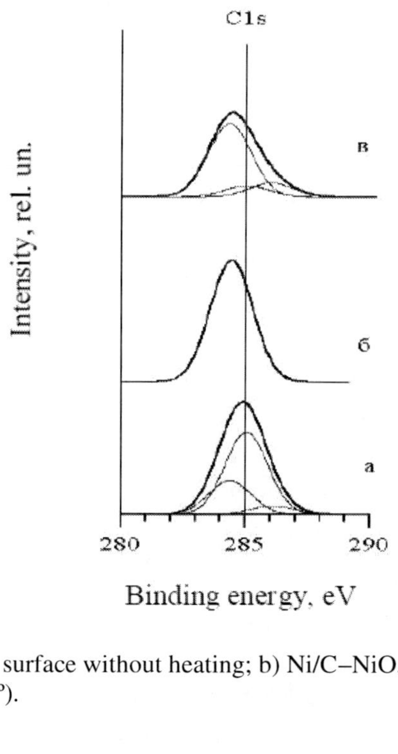


Figure 8. C1s spectra: a) surface without heating; b) Ni/C–NiO, 1:1 (300ºC);
c) Ni/C–NiO, 1:0,5 (300º).

**Table 7. Parameters of multiple splintering Me3s spectra in Ni/C nanocomposite (Ni/C NC) and in systems, containing Ni-C MC – NiO**

| Sample | $I_2/I_{1Ni}$ | $\Delta_{Ni}$, eV | $\mu_{Ni}$,$\mu_Б$ |
|---|---|---|---|
| Ni-C MC | 0.32 | 3,0 | 1.8 |
| Ni-C MC+ NiO= 1:0,5 | 0.6 | 2.6 | 3.0 |
| Ni-C MC+ NiO= 1:1 | 0.8 | 2.2 | 4.0 |

$I_2/I_1$ – the relation of multiple splintering lines maximums intensities;
Δ - Energetic distance between the multiple splintering maximums in Me3s spectra.

The Oxygen content is more much decreased at the sample heating. The C1s spectra (Figure 8) consist from three components which take place [2] in Carbon nanostructures: 285 eV(C–H bond); 284 eV ($sp^2$ hybridization); 286 eV ($sp^3$ hybridization).

According to C1s spectra for system "Ni-C MC – NiO" (relation 1:1) the maximum intensity of $sp^2$ hybridization peak is only remained at the sample heating to 300ºC (Figure 7b), and the C–H component intensity for the sample with relation 1:0,5 is decreased as well as the correspondent intensity at the relation 1:1 becomes equaled to zero. These changes are accompanied by changes of Nickel atomic magnetic moments:

- for system of content 1 Ni-C MC–0,5 NiO the Ni atomic magnetic moment increasing corresponds to 1,2 $\mu_B$;
- for content system 1 Ni-C MC–1NiO analogous increasing value equals to 2,2 $\mu_B$ (Table 7).

It's possible that the increasing of Nickel atomic magnetic moments proceeds because of the growth of atomic magnetic moments both mesoparticle cluster metal and modifier (NiO) metal.

The distance decrease between the maximums of multiple spintering in Ni3s spectra for systems shows on the increase the hybridization of d electrons in Nickel atoms–that is the reason of Nickel atomic magnetic moment growth.

Thus, when the mesoparticles modification reactions with the using APP, $SiO_2$, metal oxides are carried out, the redox processes are realized. In these cases the modifiers reduction reactions takes place. These processes are accompanied by the metal atomic magnetic moments growth, that is explained by the electron shift on high energetic levels because of the annihilation phenomenon. The energy which is evolved at this can be used in chemical rections including connection to mesoparticle. On the base of studies results the hypothesis of possibility of two phenomena (interference and annihilation) in redox processes is proposed. The correspondence of results obtained by the methods of x-ray photoelectron spectroscopy and electron paramagnetic resonance (EPR) is given. The of changes for metal atomic magnetic moments in dependence on the number of electrons which participated in redox process is observed.

## 3.2. Reagents and Catalysts in Radical Processes

According to [40] in the presence of chemical particles containing unpaired electrons (or radicals) and double bonds the addition reactions flow on the radical mechanism. Below the reactions between Copper Carbon mesoscopic composite and organic amines or halides are considered. The explanation of these reactions mechanism concludes the notions of negative charges quantization and interference. The quants from unpaired electrons are active agents interacted with reagents across the formation radical particles which are recombining. The examples of these reactions with the participating of polyethylene polyamine (PEPA) and ammonium iodide (AmI) are presented below.

The interaction of Copper Carbon mesocomposite with polyethylene polyamine (PEPA) and also ammonium iodide (AmI) is studied by IR and x-ray photoelectron spectroscopy. In these cases according to spectra the nitrogen and iodine addition to carbon shell without change of oxidation states as well as the absence of metal atomic magnetic moment changes in mesoparticles is observed.

The modification processes of Copper Carbon mesocomposite is carried out by means of the grinding of mesoparticles with reagents (PEPA or AmI) in mechanical mortar with the duration equaled to 3 minutes (the energy used corresponds to 220 kJ/mol, approximately).

It's possible to suppose the following stages:

- Mechanical action on the mixture of mesoparticle and reagent initiates unpaired electrons on mesoparticle carbon shell and negative charge quants begin to move in side to positive charged atom of reagent.
- Then the increasing of reagent polarization with the negative charge quants flow formation proceeds, and farther the negative charge quants move to opposite flow of mesoparticle negative charge flow.
- These processes lead to the interference and the formation of such chemical bonds as C–N and C–I bonds.
- In parallels the Hydrogen from reagent (for instance, PEPA) is added to carbine bonds. Therefore, at the amine group interaction with mesoparticle carbon shell the formation of C=N bond is possible.

The proposed scheme is confirmed by the results of C1s and N1s spectra (Figure 9, 10). In C1s spectrum of Cu/C nanocomposite modified by PEPA the following components as C–H (285 eV), C–N (286,4 eV) and C=N (289 eV) are found. It's necessary to note that the peak at 289 eV has the small intesity. This fact is explained by the small quantity of $NH_2$ groups in PEPA because these groups are the ultimate groups in olygomer.

In N1s spectrum of Cu-C mesocomposite (Figure 10) the presence of C–N (398,8 eV) and N–H (397 eV) bonds is observed.

The results of C1s and N1s spectra are confirmed the data of IR spectroscopy for samples modified by PEPA.

In IR spectra the additional peaks at 1075 and 1268 $cm^{-1}$ are appeared. These peaks can be relevant to C–N bonds in Nitrogen containing compounds including bonds in the adsorbed PEPA on the surface of Copper Carbon mesocomposite.

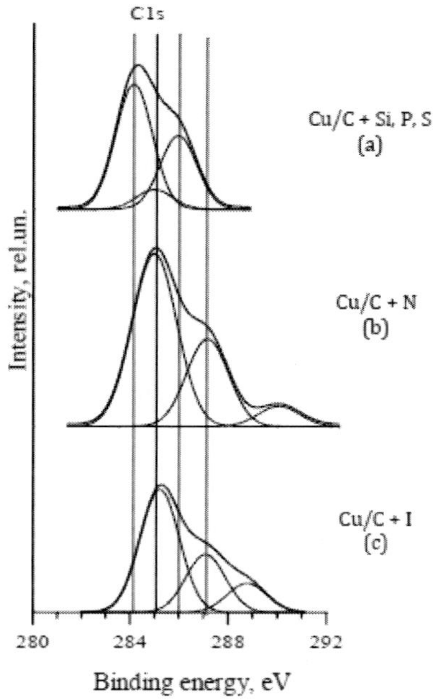


Figure 9. C1s spectra of Cu/C nanocomposite moified by PEPA (a) and by AmI (b).

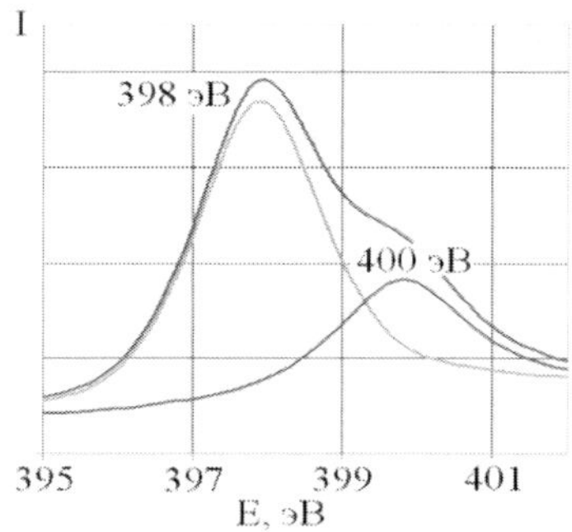


Figure 10. N1s spectra of Cu/C mesocomposite modified by PEPA.

Thus, in the case, when in the reactions with mesocomposite such reagents as polyethylene polyamine and ammonium iodide are applied, the connection reactions takes place. As the main method in the investigations x-ray photoelectron spectroscopy is applied. Owing to this method the chemical bonds formation between mesoparticles carbon shells and

Nitrogen or Iodine is found because of the interference of negative charge quants during the interaction processes of polyethylene polyamine or ammonium iodide with mesocomposite. At the interactions of PEPA with mesoparticles the C–N bond formation is explained by the interference of negative charges quants.

### 3.3. Reactivity of Metal Carbon Mesocomposites in the Coordination Reactions and at the Metals Sorption

The metal carbon mesoscopic composites reactivity in the coordination reactions is caused by the electron structure of these composites which also determines the nanostructures forms and the peculiarities of its interactions with media or reagents. It's necessary to note that the metals of mesocomposites are coordinated on carbine or on poly acetylene fragments of these nanostructures carbon shells. The coordination reaction with the carbine fragment of carbon shell is more probable because of great electron density. The presence of metal cluster coordination to carbon shell secures the electron transport from metal to carbon shell fragment. In this case the possibility of reagents metals coordination is appeared.

For these cases the following coordination processes mechanism can be proposed:

1. The negative charge quants stream is directed to the electron shell of reagent atom (may be metal) and activates the electron shell quants.
2. This activation can to initiate the positive charged quants appearance and then the annihilation phenomenon.
3. In order to the annihilation stimulates the interference process development which provides the metal sorption from correspondent medium.

If the metal oxide is applied as modifier agent, it's possible the competition of metal reduction processes between modifier agent metal and mesocomposite (or mesoscopic granule) metal. For instance, at the Cu-C MC modification by Nickel oxide it's observed the atomic magnetic moments growth for Copper (cluster metal within mesogranule) on 0,7 $\mu_B$, and for Nickel – 0,5 $\mu_B$ (Table 8).

**Table 8. Parameters of multiplet splitting of Ni3s and Cu3s spectra for Cu-C mesocomposites modified by Nickel oxide, including atomic magnetic moments of metals**

| **Sample** | **$I_2/I_1$** | **Δ, eV** | **$\mu_{met}$, $\mu_B$** |
|---|---|---|---|
| Cu3s $_{nano}$ (Ni) | $0{,}4_{Cu}/0{,}4_{Ni}$ | $3_{Cu}/2_{Ni}$ | $2_{Cu}/2{,}3_{Ni}$ |

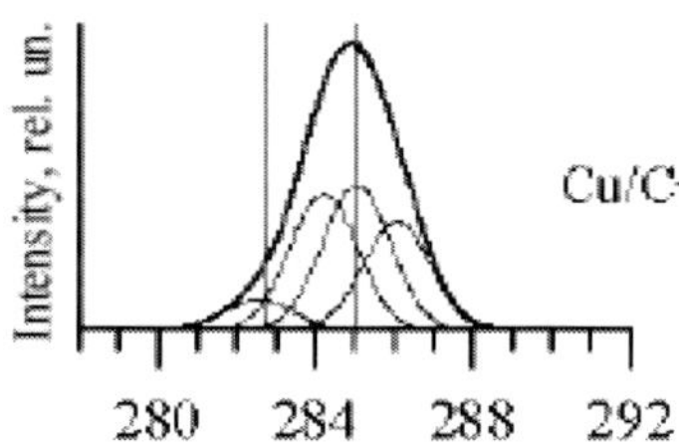


Figure 11. C1s spectrum of Nickel containing of Copper Carbon mesocomposite.

**Table 9. Atomic magnetic moments of metals in modified Nickel Carbon mesocomposite**

| **Mesocomposite and Modifier (mass part)** | **$\mu_{Ni}$, $\mu_B$** | **$\mu_{Cu}$, $\mu_B$** | **$\mu_{Fe}$, $\mu_B$** |
|---|---|---|---|
| Ni-C MC – NiO (0,5) | 3,0 | | |
| Ni-C MC – NiO | 4,5 | | |
| Ni-C MC – CuO | 2,0 | 2,0 | |
| Ni-C MC – $Fe_2O_3$ | 2,5 | | 3,0 |
| Ni-C MC – $Al_2O_3$ (0,2) | 4,8 | | |

Also the deformation of carbon shell takes place as well as the sp hybridization appearance (carbine fragments) (Figure 11).

The metal atomic magnetic moment changes at Nickel Carbon mesocomposite modification by the following metal oxides as CuO, NiO, $Fe_2O_3$, $Al_2O_3$ are adduced in Table 9.

On the base of results of Ni-C mesocomposite modification it's possible to conclude that the ability to polarization of metal oxide layer influences on the Nickel atomic magnetic moment growth. For example, the significant growth of Nickel magnetic moment takes place if the Aluminum Oxide is used as modifier. During this process the Aluminum is completely reduced (according to data of x-ray photoelectron spectroscopy).

At the beginning of processes during the joint grinding of components the multiples (mesoscopic reactors) are formed and simultaneously the flows of quants of negative and positive charges are appeared. In this case there is phenomenon of annihilation with the formation of electromagnetic field which stimulates the atomic magnetic moment growth.

# 4. Metal Carbon Mesocomposites Application Possibilities as Magnetic Mesoscopic Materials

## 4.1. The Glues and Adhesives on the Base of Metal Carbon Mesoscopic Composites

The presence of unpaired electrons and double bonds in carbon shell of metal carbon mesocomposites guarantees the additional conditions for the adhesion increasing especially at the metal materials connection. In these cases the positive meaning has the presence of above mesocomposites magnetic properties. For example, the introduction of Nickel Carbon mesocomposite in the composition of Silver containing current conductive

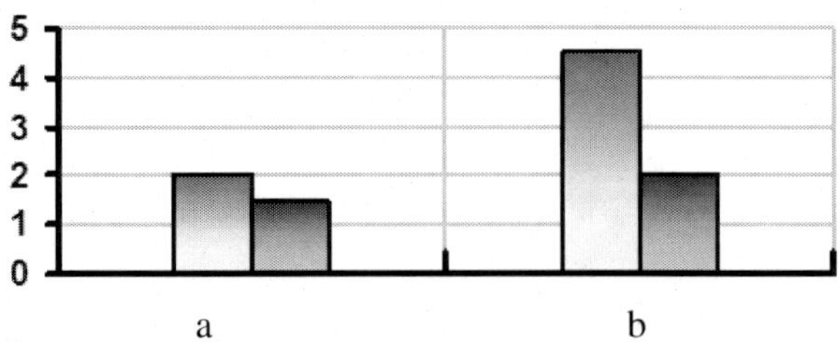


Figure 12. The adhesion durability on shift for Ag containing glue (pink) and paste (blue), initial (a) and modified by Nickel Carbon mesocomposite.

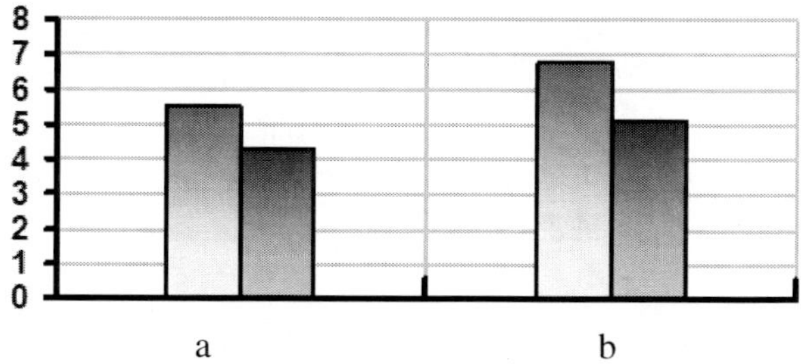


Figure 13. The adhesion durability on separation for Ag containing glue (pink) and paste (blue), initial (a) and modified by Nickel Carbon mesocomposite.

**Table 10. The measuring results of the current conductive paste (glue) electro resistance, modified by Nickel Carbon mesocomposite**

| **Characteristics** | **Current conductive paste** | **Current conductive glue** |
|---|---|---|
| Specific volume electro resistance, Om·cm (*initial sample*) | $2{,}4\cdot10^{-4}$ | $3{,}6\cdot10^{-4}$ |
| Specific volume electro resistance, Om·cm *(modified sample)* | $2{,}2\cdot10^{-5}$ | $3{,}3\cdot10^{-5}$ |

glue leads to the improvement of adhesion and electric conductivity (Figure 12, 13; Table 10).

Thus, the Metal Carbon mesocomposites can be applied for the improvement of the current conductive materials characteristics.

Analogous results are obtained at the modification of cold hardened epoxy resins by the Metal Carbon mesocomposites. According to the investigation on the modification results of cold hardened epoxy resins the following conclusion may be made:

> The test for defining the adhesive strength and thermal stability correlate with the data of quantum-chemical calculations and indicate the formation of a new phase facilitating the growth of cross-links number in polymer grid when the concentration of Cu-C mesocomposite goes up. The optimal concentration for elevating the modified epoxy resin (ER) adhesion equals 0.003% from ER weight. At this concentration the strength growth is 26.8%. At the same time the optimal quantity of Cu-C

mesocomposite for elevating the modified industrial epoxy materials adhesion equals 0,005% that leads to the strength growth equals 60.7%. From the concentration range studied, the concentration 0.05% from ER weight is optimal to reach a high thermal stability. At this concentration the temperature of thermal destruction beginning increases up to 195°C.

The modification of hot hardened epoxy resins by means of Copper Carbon mesocomposites is carried out with the application of the finely dispersed suspension based on iso methyl tetra phtalic anhydrate or based on toluene. After testing the samples of four different schemes, the increase in the strength at detachment $\sigma_{det}$ up to 50% and shear $\tau_{sh}$ up to 80% takes place, the concentration of Copper Carbon mesocomposite introduced corresponds to 0.0001-0.0003%.

The application of these materials as adhesives for the gluing of metals and vulcanite is realized on the schemes "metal$_1$ – adhesive$_1$ – vulcanite – adhesive$_2$ – metal$_2$." To define the adhesive tear and shear strengths the above proposed scheme was used (Figure 14, 15). The investigations carried out revealed that the modification of the conventional recipe of the glue 51-K-45 results not only in increasing the glue adhesive characteristics but also in changing the decomposition character from adhesive-cohesive to cohesive one.

The availability of metal compounds in mesocomposites can provide the final material with additional characteristics, such as magnetic susceptibility and electric conductivity.

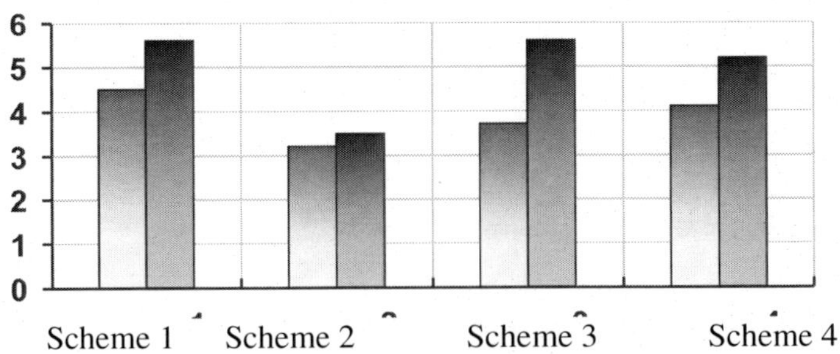


Figure 14. Relative changes of adhesive tear strengths of epoxy glues modified by Copper Carbon mesocomposites (content of MC – 0,0001%).

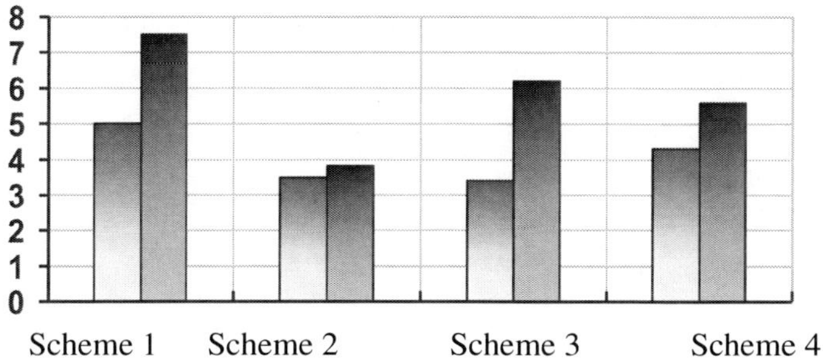


Figure 15. Relative changes of adhesive shear strengths of epoxy glues modified by Copper Carbon mesocomposites (content of MC – 0,0001%).

The modification of different materials by minute quantities of Metal Carbon mesocomposite allows improving their technical characteristics, decreasing material consumption and extending their application.

## 4.2. Magnetic Transport of Medicines within Organism from Metal Carbon Mesocomposites

At last time the great attention is spared to the creation of medicine remedies with address direction for the action on the organs which are needed in correspondent healing. Usually the transport of therapeutically active substances to the correspondent organ is carried out by means of the magnetic mesoparticles containing the linker connected with medicine substance. In other words, the remedies of medicine carriage in organisms are constructed on the following scheme: transport magnetic mesoparticle – linker – medicine [42]. In above patents the iron containing nanosized particles, for example, $Fe_3O_4$, are used as magnetic mesoparticle, and the organic substances connected with this mesoparticle by covalent or coordinative bonds as linker. In this case it's possible the certain difficulties on the undoing of medicine because of the bio active substances big interaction with the some functional groups. The medicine release from magnetic mesoparticles with linker occurs at the variable magnetic fields. The best linker can be phosphorus organic compounds which, as it's known, are easily destructed in water media. Therefore it's proposed [34] to accomplish the modification of metal containing magnetic mesoparticle by ammonium polyphosphate. At the same time the copper carbon mesocomposite [37] is proposed as magnetic mesoparticle since the Copper has bactericides' and anti microbes' properties and increases the organism protective forces. For that reason the Copper Carbon mesoscopic composite modified by ammonium polyphosphate chooses as the investigations object. The modification process is carried out by mechanic chemical method in processing which the Phosphorus reduction and the formation of linker which consists phosphorus with the following oxidation states as 0, +3, +5 takes place.

The obtaining of remedy for medicine delivery to definite organ is carried out on the following scheme:

> The phosphorus containing copper carbon mesoscopic composites ($MC_{(P)}$) are applied as magnetic mesoparticles. Then the medicines (M) are linked with $MC_{(P)}$ at the relation ($MC_{(P)}$ /M) correspondent to 1:(0,02–0,5) by means of mechanic chemical method with the using the mechanical mortar. The $MC_{(P)}$ applied are obtained by the interaction of copper carbon mesocomposite (Cu-C MC) and ammonium polyphosphate (APP) at the relation 1:0,5 for the obtaining of high atomic magnetic moment of Copper. The therapeutically active substances such as adenosine tri phosphorus acid (the relation $MC_{(P)}$ /M = 1:0,02), ascorbic acid ($MC_{(P)}$ /M = 1:0,2) and Urotropine ($MC_{(P)}$ /M = 1:0,5) is linked with phosphorus containing mesocomposite across the phosphorus containing liker.

The investigations of phosphorus containing Copper Carbon mesocomposite and its analogous are realized with the application of methods complex from which the basic methods are x-ray photoelectron spectroscopy (RPES), transition electron microscopy of high solution (permeation) (TEM), electron micro diffraction, electron paramagnetic resonance (EPR), x-ray diffraction measuring.

According to these investigations the copper atomic magnetic moment growth is established at the copper carbon mesocomposite modification by ammonium polyphosphate and therapeutically active substances. The atomic magnetic moment of Copper is obtained more ($\mu$ = 4,5$\mu_Б$) in the comparison with iron containing nanostructures ($\mu = 2{,}5\mu_Б$). Hence the proposed magnetic mesoparticles can be interested as the remedies of medicine carriage in organisms by means of drive magnetic field.

The production of magnetic mesoscopic particles with connected therapeutically active substances is realized by mechanic chemical method on the next scheme:

- The reduction-oxidation synthesis of copper carbon mesocomposite from CuO and polyvinyl alcohol at the reagents relation equaled to 1 mol : 4 mol [31]
- The copper carbon mesocomposite modification by ammonium polyphosphate at the reagents relation [33, 39]
- The phosphorus containing copper carbon mesocomposite modification by therapeutically active substances, such, as adenosine tri phosphoric acid (relation – 1:0,02), ascorbic acid (relation – 1:0,2) and Urotropine (relation – 1:0,5). (Pat. 2018143197).

The mechanic chemical synthesis is carried out with the using of mechanical mortar by the joint grinding of reagents at the energetic expenses approximately equaled to 260–270 kJ/mol. After mechanic chemical process the mesoscopic product obtained is dried in the closed crucible at 400ºC for the first stage, and at 150ºC for second and third stages. Then the product obtained is standee in vacuum at 100–150ºC during 3 minutes.

The results of mechanic chemical process with thermo chemical finishing are estimated with the application of following methods: x-ray photoelectron spectroscopy, electron paramagnetic resonance, transition electron microscopy with high permission.

The transition analytic electron microscope FET Tecnai G2F20 with prefix EDAX is used for the investigation of copper carbon mesoscopic composite structure and phase content. High permission corresponds to 2nm in 1,5 cm.

X-ray photoelectron spectroscopic investigations are realized by x-ray photoelectron magnetic spectrometer with the permission $10^{-4}$ at the activation $AlK_{\alpha}$ line 1486 eV in vacuum $10^{-8}$–$10^{-10}$ Torre. On the basis of Van Fleck theory the model for metal atomic magnetic moments calculation is proposed.

EPR investigations are carried out by means of EPR spectrometer E–3 of firm «Varian».

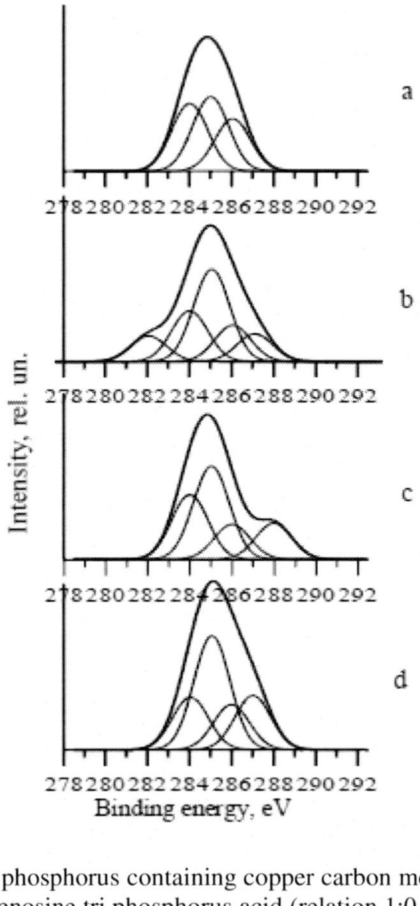


Figure 16. C1s spectra phosphorus containing copper carbon mesocomposite (a) and its analogous modified adenosine tri phosphorus acid (relation 1:0,02) (b), ascorbic acid (relation 1:0,2) (c), Urotropine (relation 1:0,5) (d).

The peak correspondent to binding energy 132,5 eV is ascribed [43] bond C=P that can be possible at the appearance of interference phenomenon because of the direct electromagnetic field which arises at annihilation in red-ox process. The unpaired electrons presence on the carbon shell of mesocomposite can promote the decreasing of free radical activity in the defeat part of organism. The high atomic magnetic moment of Copper (more than 4$\mu_B$) can be used for the address carriage of medicine.

Thus, the transport remedy included the copper carbon mesocomposite and linker containing phosphorus for the connection therapeutically active substances is proposed.

The investigations of medicines modified by above magnetic mesocomposites are carried out with the application of x-ray photoelectron spectroscopy. Below C1s spectra for these medicines are presented (Figure 16).

The discussion of the electron structures changes for the modified bio active substances takes place below in following order:.1) ATP; 2) AA; 3) U.

After the modification of phosphorus containing copper carbon mesocomposite, obtained at the relation Cu-C MC/APP equaled to 2, by adenosine tri phosphorus acid (relation 1:0,02) the following changes of the mesoparticle surface structure takes place:

- In C1s spectrum the maximum for C–H bond is considerably increased;
- In this spectrum the appearance new content at 283 eV correspondent to sp hybridization, and the maximum for $sp^2$ (C–C) and also $sp^3$ (C–C) hybridization are decreased;
- In C1s spectrum new maximum (286,7 eV) for C–N bond is observed;
- In Cu2p and P2p spectra the peaks for Cu–P and Cu–P–O are found.

It's necessary to note that the magnetic moment determined on Cu3s spectrum did not exchange and is equaled to 4,5$\mu_B$.

Above changes in spectra can be explained by the carbon phosphorus shell deformation at the interaction of adenosine tri phosphorus acid with transport remedy.

In the x-ray photoelectron spectra of mesoparticles obtained by means of the phosphorus containing mesocomposite modification by ascorbic acid at the reagents relation 1:0,2 the following changes are discovered:

- In C1s spectrum in comparison on the spectrum of initial mesocomposite the new maximum ($E_b$ = 288 eV) for C–OH group is present;
- In Cu2p and P2p spectra the maximums for Cu–P, Cu–O and P–O bonds are observed.

The magnetic moment determined on Cu3s spectrum also as in above example is not changed and equaled to 4,5$\mu_B$.

The changes in x-ray photoelectron spectra of phosphorus containing Cu-C mesocomposite modified by Urotropine at the relation 1:0,5 also take place. Below the following peculiarities in spectra are given:

- In C1s spectrum the C–H maximum increasing is found;
- In the same spectrum the maximum for C–N bond ($E_b$ = 287eV) is appeared, and the maximums attributed to $sp^2$ and $sp^3$ hybridization have near intensity;
- In P2p spectrum the maximums at 135 eV ($P^{+5}$) and 129 eV ($P^0$) take place.
- In N1s spectrum the covalent bond C–N is determined.

In this case the Copper atomic magnetic moment is near to 4,5$\mu_B$.

It's interest to note that the carbine component (sp hybridization) is observed at the mesocomposite quantity increasing.

Thus, the above said examples show on the connection of therapeutically active substances with transport remedy (phosphorus containing copper carbon mesocomposite) at the conservation of magnetic properties. Latter is necessary for the application of the obtained substances in practice.

## 4.3. Stimulators for Plants Growth with Additives of Metal Carbon Mesocomposites

The Metal Carbon mesoscopic composites owing to structure and properties can be effective for the increasing and improvement of agricultural producing. These substances can be initiators or inhibitors for the definite vital processes in different fields of agriculture. In this paper as examples the stimulation of plants growth under the action of Metal Carbon mesocomposites are considered. As examples of plants the lilies are discussed. In this case the bulbs of lilies came in peat; they were packed in film bags, from a dense film that had 18 holes 1 cm in diameter for air intake. Packages with lily bulbs were stored in container boxes at a temperature of -1 °C, which allows them to remain at rest. Samples of this peat were sampled and agrochemical analysis was carried out using a volumetric analysis of the preparation of aqueous extract (with recalculation of the results of the analysis for volume) and a sample with recalculation of the results of the analysis for dry weight [35] (Table 11).

**Table 11. Agrochemical characteristics of peat during storage of lily bulbs imported from Holland**

| Index | Result | The value of the horse Peat sphagnum |
|---|---|---|
| Moisture of peat, % | 60,0 | 59,46 |
| pH (suspension) | 4,0 | 5,5–6,1 |
| Specific conductivity of the suspension, mS/cm, at 25°C | 0,093 | - |
| The content of mobile forms of batteries | | |
| Nitrate nitrogen $N\text{-}NO_3$ | 0,57<br>0,1 | 26,3<br>29,0 |
| Nitrogen ammonium $N\text{-}NH_4$ | 7,36<br>1,5 | 135,6<br>149,0 |
| $P_2O_5$ | 1,64<br>0,3 | 291,0<br>320,0 |
| $K_2O$ | 34,56<br>7,0 | 353,0<br>388,0 |
| $Ca^{2+} + Mg^{2+}$, mmol/g | 2,98<br>0,6 | - |

Numerator (for dry matter) - mg/100 g, denominator - mg/l.

The moisture content of the peat corresponds to the normative indices. According to the degree of acidity, peat refers to a very acidic. The specific electrical conductivity of peat does not exceed 1,0 mS/cm. Peat has a low content of nitrogen, phosphorus and potassium in terms of the degree of supply of food elements. Such properties of this peat contribute to the long storage and transportation of bulbs of long lilies and do not allow them to germinate.

For the growth of plants on top peat, the optimum moisture content is within 78-85% of the mass. The moisture content in the peat mixture is normal.

A very important factor for the development of roots and the absorption of nutrients by them is the acidity of the soil. High acidity of soil leads to insufficient intake of such elements as phosphorus, magnesium and iron. When growing Oriental hybrids of lilies, which include our variety, the pH should be between 5,0 and 6,5.

Lilies are sensitive to soil salinity. With a high content of salts in the soil, the roots of the lilies become hard, brittle, and acquire a yellow-brown color. Soil salinity is determined by the specific electrical conductivity. Lilies require low EU levels.

The content of mobile forms of nutrients is low, but the level of plant nutrition is regulated by a computer program in accordance with the needs of lily bulbs. Thanks to the inherent peat strength of buffer and high sorption capacity, mineral fertilizers are not washed out and stored in an accessible form for plants; at the same time, the danger of creating an increased concentration of salts harmful to plants is reduced.

High productivity and quality of flowers largely depend on the full provision of plants with trace elements, so when planting lily bulbs, an analysis was carried out for the content of trace elements in the peat mix (Table 12).

According to the content of trace elements, peat mixes contain a very low content of such elements as Zinc, Copper, Iron, Manganese and Boron [8]. By processing bulbs of lilies, Copper Carbon mesocomposite, we contribute to replenishing available forms of Copper in lily plants. In the period of growth and development, the level of nutrition of lily plants is regulated by a computer program in accordance with the needs of macro- and microelements.

For two years, the reaction of the lily to the treatment of Copper Carbon mesocomposite was studied. The biometric parameters of the lily were determined as the height of the flower bud in the bud budding phase (Table 13), the number of buds, the diameter of the open flower, and the height of the stem when cutting plants.

The number of buds on the shoot was higher for the Siberia variety, than for the Santander variety by 1,1 pcs. (at HSR = 0,06 pcs.). When processing with mesocomposites, there was a significant increase in the number of buds to 5,0-5,3 pieces (Table 14).

**Table 12. Chemical analysis of peat mixture for micronutrients content**

| Microelements, (mg/kg dry weight) | The actual value of the test results |
|---|---|
| Zinc | 15,60 |
| Copper | 3,81 |
| Iron | 126,60 |
| Manganese | 57,30 |
| Boron | 8,27 |

**Table 13. Height of flowering shoot, cm (average for 2015-2016, the phase of the beginning of budding)**

| Hybrids of lilies (factor A) | Concentration of mesocomposites (factor B) | | | | | Average by factor A $HSR_{05}$ A = 1,7 |
|---|---|---|---|---|---|---|
| | Without processing (k) | Water (к) | 0,01 % | 0,02 % | 0,05 % | |
| Siberia (к) | 79,6 | 71,9 | 87,7 | 87,9 | 87,2 | 82,8 |
| Santander | 77,9 | 81,6 | 89,1 | 82,2 | 82,3 | 82,6 |
| Average by factor B $HSR_{05}$ B = 2,7 | 79,2 | 76,9 | 88,5 | 85,2 | 84,6 | - |
| $HSR_{05\ \text{private differences}}$ | 3,8 | | | | | |

**Table 14. Number of buds, pcs (average for 2015-2016, the phase of the beginning of budding)**

| Hybrids of lilies (factor A) | Concentration of mesocomposites (factor B) | | | | | Average by factor A $HSR_{05}$ A = 0,06 |
|---|---|---|---|---|---|---|
| | Without processing (k) | Water | 0,01% | 0,02% | 0,05% | |
| Siberia (к) | 4,7 | 4,7 | 5,8 | 5,6 | 5,4 | 5,2 |
| Santander | 3,5 | 3,5 | 4,7 | 4,5 | 4,6 | 4,1 |
| Average by factor B $HSR_{05}$ B = 0,1 | 4,1 | 4,1 | 5,3 | 5,1 | 5,0 | - |
| $HSR_{05\ \text{private differences}}$ | 0,1 | | | | | |

The length of the bud in Siberia variety averaged 10,5 cm, while the Santander variety was 1,2 cm lower (Table 15). The largest length was found in lilies treated with a 0,01% Copper Carbon mesocomposite suspension.

**Table 15. Length of bud, cm (average for 2015-2016, the phase of the beginning of budding)**

| Hybrids of lilies (factor A) | Concentration of mesocomposites (factor B) | | | | | Average by factor A $HSR_{05}$ A = 0,06 |
|---|---|---|---|---|---|---|
| | Without processing (k) | Water (к) | 0,01% | 0,02% | 0,05% | |
| Siberia (к) | 10,1 | 10,3 | 10,9 | 10,6 | 10,6 | 10,5 |
| Santander | 9,0 | 9,1 | 9,6 | 9,5 | 9,3 | 9,3 |
| Average by factor B $HSR_{05}$ B = 0,09 | 9,5 | 9,7 | 10,3 | 10,0 | 9,9 | - |
| $HSR_{05\ \text{private differences}}$ | 0,1 | | | | | |

The length of the bud and the number of buds depends on the photosynthetic activity, which is directly related to the width of the leaves, so the width of the lily leaf was measured during the bud budding phase (Table 16).

The variety Santander leaves was wider than that of Siberia, which is one of the varietal traits. The increase in the number and length of buds in both varieties when processing bulbs with Copper Carbon mesocomposite was associated with an increase in the width of the leaves.

### Table 16. The width of the leaf, cm (average for 2015-2016, the phase of the beginning of budding)

| Hybrids of lilies (factor A) | Concentration of mesocomposites (factor B) | | | | | Average by factor A $HSR_{05}$ A = 0,05 |
|---|---|---|---|---|---|---|
| | Without processing (k) | Water (к) | 0,01% | 0,02% | 0,05% | |
| Siberia (к) | 3,0 | 3,1 | 3,4 | 3,3 | 3,1 | 3,2 |
| Santander | 3,4 | 3,5 | 3,7 | 3,6 | 3,6 | 3,6 |
| Average by factor B $HSR_{05}$ B = 0,08 | 3,2 | 3,3 | 3,6 | 3,5 | 3.4 | - |
| $HSR_{05 \text{ private differences}}$ | 0,1 | | | | | |

### Table 17. Number of buds when cutting plants, pcs (average for 2015-2016)

| Hybrids of lilies (factor A) | Concentration of mesocomposites (factor B) | | | | | Average by factor A $HSR_{05}$=0,04 |
|---|---|---|---|---|---|---|
| | Without processing (k) | Water (к) | 0,01% | 0,02 % | 0,05 % | |
| Siberia (к) | 4,8 | 4,7 | 6,0 | 5,6 | 5,5 | 5,3 |
| Santander | 3,6 | 3,5 | 4,7 | 4,5 | 4,6 | 4,2 |
| Average by factor B $HSR_{05}$ B = 0,06 | 4,2 | 4,1 | 5,4 | 5,0 | 5,0 | - |
| $HSR_{05 \text{ private differences}}$ | 0,09 | | | | | |

### Table 18. Diameter of the open flower, cm (average for 2015-2016, cutting plants)

| Hybrids of lilies (factor A) | Concentration of mesocomposites (factor B) | | | | | Average by factor A $HSR_{05}$ A = 0,3 |
|---|---|---|---|---|---|---|
| | Without processing (k) | Water (к) | 0,01 % | 0,02 % | 0,05 % | |
| Siberia (к) | 18,7 | 18,3 | 21,3 | 20,0 | 20,0 | 19,7 |
| Santander | 20,0 | 20,0 | 22,7 | 22,0 | 21,7 | 21,3 |
| Average by factor B $HSR_{05}$ B = 0,5 | 19,3 | 19,2 | 22,0 | 21,0 | 20,9 | - |
| $HSR_{05 \text{ private differences}}$ | 0,7 | | | | | |

The productivity of the lily (lily cut) is determined by the number of buds in the inflorescence, the diameter of the flower and the height of the plant. In this regard, the plants of lilies count the number of buds when cutting plants (Table 17).

When cutting lilies, the number of buds on the plant remains the same as in the bud budding phase. There were also a greater number of buds in the Siberia lily variety than in Santander. Depending on the treatment with various concentrations, the maximum number of buds was noted at 0,01% concentration of Copper Carbon mesocomposite.

The diameter of the open flower in the Santander variety was significantly larger than that of the Siberia variety (21,3 cm) (Table 18). Processing of bulbs with mesocomposites promoted a substantial increase in the diameter of the open flower. The largest diameter of the flower was noted when treated with 0,01% mesocomposite (22,0 cm).

The height of the stem when cutting lilies increased by treatment with Copper Carbon mesocomposite to 131 - 141 cm in comparison with the control (99,0-102,3 cm). The sort of lily Siberia was slightly higher than the Santander variety.

Thus, the treatment of bulb lily Copper Carbon mesocomposites contributes to the increase in plant height, the number and diameter of the flowers.

According to the results of scientific and industrial experiments, processing of bulb lily Copper Carbon mesocomposites promotes an increase in the height of the flower sprout, the height of the stem during the cutting of plants, the number of buds and the diameter of the open flowers.

The use of 0,01% Copper Carbon mesocomposite in the cultivation of lilies led to an increase in the cost of sales and, accordingly, profit and profitability of production. Treatment of lily bulbs with a 0,01% Copper Carbon mesocomposite is energetically effective, since allowed to get products of high quality.

The mechanism of Metal Carbon mesocomposites action on the development of plants can be caused by the following reasons:

1. the definite quantity of Copper with the introduced Copper Carbon mesocomposite;
2. the magnetic properties of mesocomposite;
3. the active electrons presence on the carbon shell of Copper Carbon mesocomposite.

It's possible these factors may be substantial for the development of vital processes in plants and also in other vital organisms.

This trend of investigations needs in the development of serious theoretical elaborations in the direction to new mesoscopic biology and medicine. In these fields the Metal Carbon mesoscopic composites application for the proceeding of vital processes, compared with the studies of mechanisms of their action, can to give new impulse in the phenomena understanding, and also in the expansion of application fields of these exceptional mesoscopic particles in different trends of biology and medicine.

However at present the widest development takes place concerning to the application of Metal Carbon mesocomposites as modifiers of different materials. Below the mesocomposites application possibilities as modifiers of inorganic and organic materials is considered.

# 5. Mesocomposites Application Possibilities as Materials Modifiers

## 5.1. Metal Carbon Mesoscopic Composites as Modifiers of Concretes, Glasses and Analogous Materials

The mesocomposites activity in the different media (materials) is changed in the dependence on polarity or polarization of their media. Therefore the modification conditions for the different materials can be differed from each other.

According to scheme of possible polarization the increasing of medium (material) density takes place owing to the regular orientation of material fragments with the creation of super molecular and crystalline structures takes place (Figure 17).

Polarization growth can be expressed as $P_{com} = \Sigma p_{fg} + p_{NC}$, (1)

where $P_{com}$ – the common (summary) polarization, $\Sigma p_{fg}$ – sum of functional groups polarizations, $p_{NC}$ – the polarization (or dipole moment) of mesocomposite.

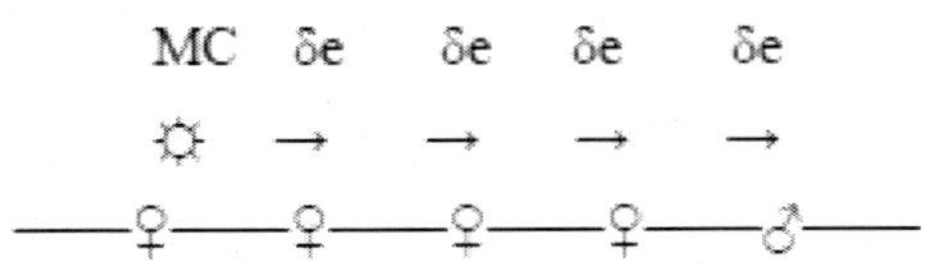


Designations: MC (☼) – mesocomposite, δe – negative charges quant (electron), → the polarization direction, ♀——♀ – macromolecule fragment with functional groups.

Figure 17. Scheme of polarization at charge quantization with expansion of quant influence on materials polar groups.

The polarization extent depends on the quants electromagnetic radiation phase velocity. It’s necessary to note that this velocity will be decreased in the media with high dielectric constant according to following formula –

$$v = c/\sqrt{\varepsilon} \quad (2)$$

where v – the phase velocity of electromagnetic radiation, c – the light velocity, ε – dielectric constant.

When the dielectric constant is increased the decrease of mesocomposite influence on the media arises and the self organization process is finished.

Depending on the development of self organization process (single measured - 1D, double measured - 2D, third measured - 3D) the super molecular structures (mesoparticles) of correspondent forms and sizes are organized. The surface energy of embryos increased influences on the mesoparticles formation. This energy can be express as the sum of energetic parts for the realization of different movements:

$$E_S = E_{(tr)} + E_{(r)} + E_{(osc)} + E_{(em)}, \quad (3)$$

where $E_S$ - surface energy of macromolecule (mesoparticle), $E_{(tr)}$ - part of translational motion energy, $E_{(r)}$ - part of rotary motion energy, $E_{(osc)}$ - part of oscillatory motion energy, $E_{(em)}$ - part of electron motion in surface layer.

In accordance with the formation of mesoparticles which have the identical orientation with each other, the parts of translational motion energy and of rotary motion energy will be near to zero. Therefore the main contribution into the mesoparticle surface energy will be bring the oscillatory motion and transport of electrons in surface layer of macromolecules (mesoparticles). Then the change of character of quants radiation wave propagation from 2D (in the surface plane) to 3D (in the space field at surface).

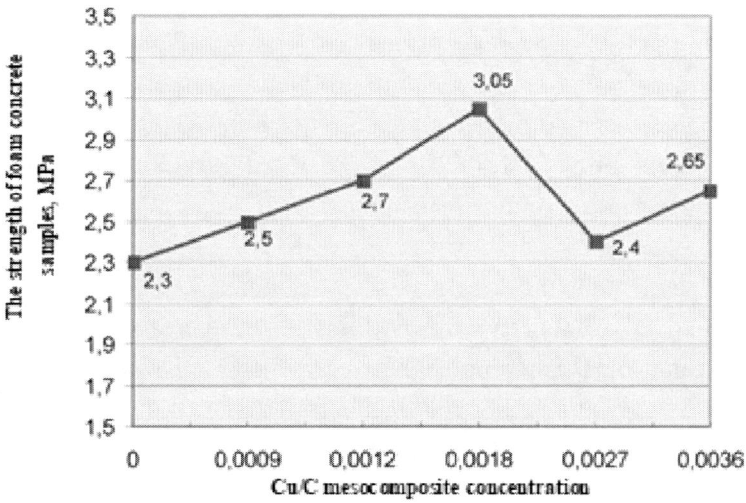


Figure 18. The dependence of foam concrete strength on Copper Carbon mesocomposite content (%).

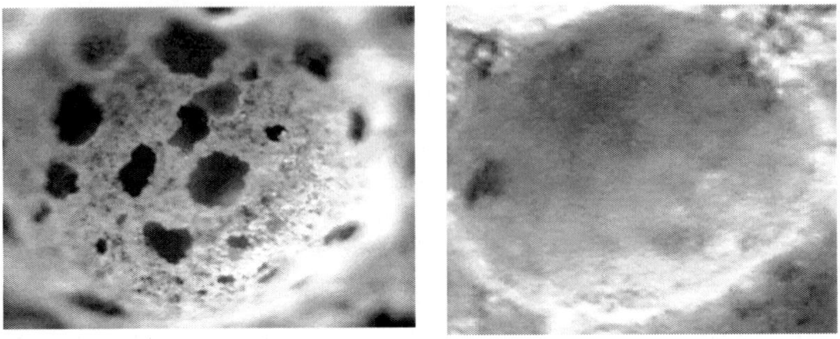

Figure 19. The foam concrete structure (×90) without mesocomposite and with the addition of mesocomposite (0,002%).

The modification of foam concrete and dense concrete is usually carried out for the increasing of compressive strength. Therefore the investigations of foam concrete strength changes in depending on content and nature of Metal Carbon mesocomposites are accomplished. Below the dependence of foam concrete strength on Cu-C mesocomposite content (in %) is given (Figure 18).

The introduction of super small quantities of mesocomposites leads to the increasing of foam concrete surface layer density (Figure 19).

The foam concrete or gas concrete production at their modification by Metal Carbon mesocomposites leads to improvement of building constructions durability in 1,5–2 times. The hard concrete production with the Metal Carbon mesocomposite using increases the strength characteristics on 30-50%.

At the same time the kinetics of these materials strength growth is studied (Figure 20). In this case the modification of foam concrete is realized with the using of mesocomposites finely dispersed water suspension. According to the results, adducted on Figure 20, the strength increasing of foam concrete modified by mesocomposite is more on 80% in comparison on foam concrete without mesocomposite. After 56 days this increasing is decreased to 30%. It is possible during the period of composition hardening the coordination of mesocomposite particle with composition molecules takes place. Analogous effect is observed during the modification of dense (hard) concrete (Figure 21).

The investigation results of kinetics of dense concrete strength growth during seven and twenty eight days is given on Figure 21. After seven days the growth of modified dense concrete strength more than the increasing of strength of non modified dense concrete on 66% and after 28 days the increasing of strength makes 43%. Owing to large density of the dense concrete composition the mesocomposite re coordination process became difficult.

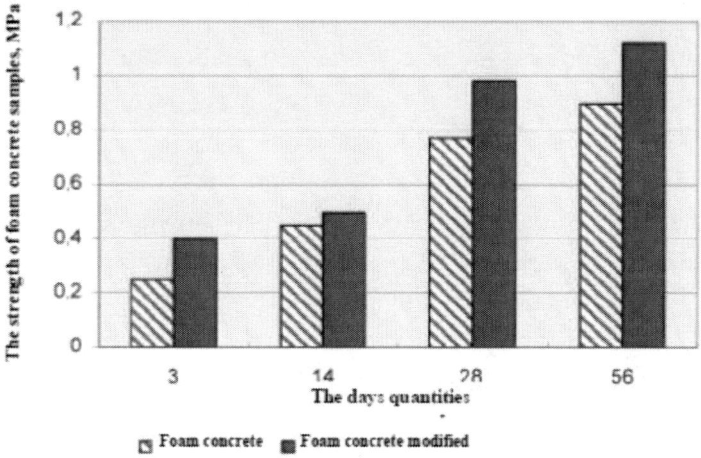


Figure 20. The kinetics of the growth of foam concrete strength.

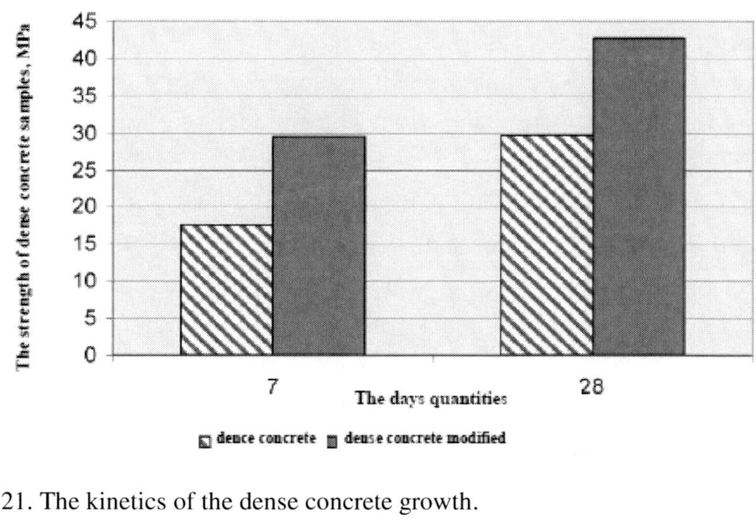


Figure 21. The kinetics of the dense concrete growth.

According to investigation results the following conclusion may be made:

1. The production of foam concrete modified by Metal Carbon Nanostructures leads to the improving of its characteristics in 1,5–2 times and to the increasing of its durability.
2. The strength characteristics of dense concrete modified by Metal Carbon mesocomposites increase on 30–50%.

The introduction of a modifier based on metal carbon mesocomposites into the composition results in medium structuring, decrease in the number of defects thus improving the material physical and mechanical characteristics.

The definite peculiarities encountered at the modification of silicates and, especially, liquid glass. The interest to liquid glass is conditioned, first of all, by its ecological friendliness and production and application simplicity, inflammability and, practically, the absence of toxicity, biological stability, and raw material availability. Such nanostructures as Metal Carbon mesocomposites can be used to improve such characteristics of paints as elasticity, adhesion to the basic construction, hydrophobic behavior and also helps solving the problems connected with coating flaking-off from the base, discoloration, limited color range, etc. Due to the unique properties of Metal Carbon mesocomposites it's possible to apply new properties to silicate paints, for example, to produce the coating protecting from electromagnetic action. The application of nanostructures in silicate materials to improve their stability to external action is known [1]. The introduction of nanostructures into the liquid glass allows qualitatively changing the material behavior in a positive way. This possibly occurs in the process of binder super molecular structure change.

Based on the aforesaid, it is important to improve operational properties of compositions on liquid glass basis modifying them with metal carbon mesocomposites, providing a wider field of their application.

Before the liquid glass modification by Metal Carbon mesocomposites the correspondent preparation of reactive composition needs. It' concluded in the ultrasound processing of reactionary masses for the fine dispersion obtaining. In this case it's necessary to define the processing time interval during which the optical density will be at the maximum, i.e., it will correspond to the maximal saturation of mesocomposite suspension. To define the time interval of ultrasound processing four suspensions were prepared for each soaking period – 3, 5, 10, 15 min, respectively, and also the reference solution without ultrasound processes – for comparison. The concentration in all solutions was 0.003%.

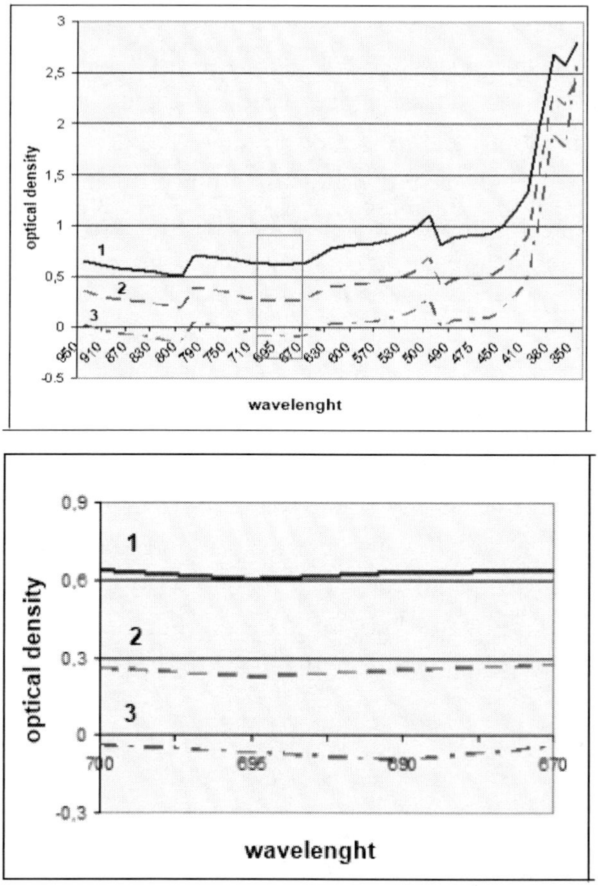


Figure 22. Changes of liquid glass (3) optical density in comparison with optical densities of liquid glasses, modified by Fe-C (1) or Ni-C (2) mesocomposites.

Thus the optimal time interval for ultrasound processing is 5 minutes. Further processing of suspensions for investigation will be carried out within this interval. After the preparations of correspondent suspensions or glue samples with Iron or Nickel containing mesocomposites the optical (visible) spectra investigations are carried out. In accordance with the results of investigations of modified films in the glue sample with Fe and glue with Ni the shift of optical density is observed at some wavelengths, indicating the changes taking place when nanostructures are introduced (Figure 22).

At other wavelength values the optical density only increases indicating the possibility to apply nanostructures as coloring pigments in paints. At the same time, the increase in the optical density of films with nanostructures in comparison with liquid glass films possibly indicates the increase in the density of compositions and formation of new structural elements in them.

Then the heat-physical properties of samples obtained are studied. At the beginning to obtain the details of changes in heat-physical characteristics, the heat capacity and thermal conductivity of the samples based on cardboard and modified liquid glass are investigated. The samples are prepared gluing several layers of cardboard with liquid glass modified with mesocomposites. The sample dimensions are found by the technique for measuring heat capacity and thermal conductivity. The sample being tested was 15 mm in diameter and 10 mm in height. The sample dimensions are measured with micrometer with 0.01 mm accuracy. The sample mass was measured with the allowance not exceeding 0.001g. Heat capacity of the samples is investigated on c-calorimeter IT-c-400. To find thermal conductivity the calorimeter IT-λ-400 is used. Further the specific thermal conductivity of the sample was calculated as follows:

$$\lambda = h/R_s, \qquad (4)$$

where λ – specific thermal conductivity, W/m*K; h – sample thickness, m.

The results of thermal physical investigations are given in Table 19.

Heat capacity of the sample, containing Fe-C mesocomposite, increased on 20% on the relation to the standard sample, and sample with Ni-C mesocomposite– on 66%. At the same time, thermal conductivity decreased on 27% and 23% for the samples with Fe and Ni, respectively. So when nanostructures are introduced, in the average the characteristics change as follows: density increases by 13%, heat capacity – by 40%, thermal conductivity decreases by 25%. Further, using the experimental results, the temperature conductivity is calculated by the following formula:

$$a = \lambda/c\rho, \tag{5}$$

where $a$ – temperature conductivity coefficient, λ – thermal conductivity coefficient, c – heat capacity, ρ – density.

Inserting the experimental data into the formula, we can define the temperature conductivity values (in percent):

$$a = \lambda/c\rho = 0.75\lambda_0/1{,}4c_0{*}1.13\rho_0 = 0.47a_0. \tag{6}$$

or calculate separately:

$$a_1/a_0 = \lambda_1\rho_0C_0/\ \lambda_0\rho_1C_1{=}0.51 \tag{7}$$

$$a_2/a_0 = \lambda_2\rho_0C_0/\ \lambda_0\rho_2C_2 = 0.43 \tag{8}$$

where $a_1$, $\lambda_1$, $\rho_1$,C1 – characteristics of the sample with Fe/C NC, $a_2$, $\lambda_2$, $\rho_2$,$C_2$ – characteristics of the sample with Ni/C NC, $a_0$, $\lambda_0$, $\rho_0$,$C_0$ – characteristics of non-modified sample.

**Table 19. Thermal physical characteristics of the samples**

| **Samples** | **cardboard/glue without Fe and Ni mesocomposites** | **cardboard/glue with Fe (change in %)** | **cardboard/glue with Ni (change in %)** |
|---|---|---|---|
| Density, kg/m$^3$ | 624.5 | 744 (↑19%) | 669 (↑7%) |
| Heat capacity $C_{spec}$, J/kg*K | 1790 | 2156 (↑20%) | 2972 (↑66%) |
| Thermal conductivity λ, W/m*K | 0.083 | 0.061 (↓27%) | 0.064 (↓23%) |

*Thus, temperature conductivity decreased by nearly 50% in comparison with the initial values ($a_0$).*

When nanostructures are introduced, self-organization takes place. Mesoparticles structure the silicate matrix leading to the formation of new elements in the structure, thus increasing the material density and influencing its heat-physical characteristics. When additional structural elements and new bonds are formed, the system internal energy increases leading to heat capacity elevation and, consequently, temperature conductivity decrease. Thermal conductivity decrease of silicate paints when applied as a coating allows improving heat-physical characteristics of the whole protective structure of a building. In turn, temperature conductivity decrease results in decreasing the amount of heat passing through the coating, thus preserving adhesive characteristics of the coating for a long time.

## 5.2. The Application of Metal Carbon Mesocomposites as Modifiers of Polymeric Materials and Polimeric Composites

When the additive sizes are decreased to nanometer sizes the phenomena of mesoscopic particles are appeared. According to [41] the phenomena such as interference, spectrum quantization, charge quantization occur when the mesoparticles have the limitation in motions or in the energetic possibilities realization. In this case the mesoparticle can only vibrate and also the electron transport is possible.

The media (or polymeric compositions) properties changes under the mesoparticles influence can be achieved at equal distribution of these particles in composition volume and at its coagulation absence. Last is possible at the following conditions:

- certain polarity and dielectric constant of medium;
- minute concentration of mesoparticles;
- ultrasound action on the correspondent suspension for the proportional distribution of mesoparticles.

The assignment of active nanostructures (mesoparticles) during the compositions modification is concluded in the activation of matrices self organization in needful direction. For the realization of this goal the determination of organized phase part is necessary. The Avrami–Kolmogorov equation is applied for the organized phase determination –

$$W = 1 - \exp(-k\tau^n), \qquad (9)$$

where W – the part of organized phase, k – the parameter defined the rate of organized phase growth, τ – the duration of organized phase growth, n – the fractal dimension.

In some papers [9–17] the positive results on materials properties improvement are presented when the minute quantities of metal carbon mesocomposites are introduced in these materials. In paper [14] the hypothesis about nanostructures influence transmission on macromolecules of polymeric matrices is proposed. This hypothesis is complied with mesoscopic physics principles which consider quantum effects at the certain conditions of mesoparticle existence.

The composition polarization is possible because of there is the charge quantization with the wave expansion on polar (functional) groups of media (for example, polymer macromolecule) (Figure 17).

The quantum charge wave expansion leads to the functional groups polarization (dipole moments) change as well as the extinction increasing. Last bring growth of peaks intensities in IR spectra. The individual peaks growth effects in IR spectra are observed at the introduction of mesocomposites minute quantities (Table 20). Let us note, that the peaks intensity growth in IR spectra is observed when the quantity of introduced mesocomposite is decreased. This fact is complied with fundamental principles of chemical mesoscopic. In illustrated case the instance of fine dispersed suspension Cu-C mesocomposite (hardener for epoxy resin). According to data of Table 3 the decreasing of mesocomposite quantity to 0,001% leads to the growth of some peaks intensity in IR spectra.

At the second day of that suspension existence the floccules are formed and peaks intensity sharply drops. However the suspension activity can be increased with the using of ultrasound treatment. The treatment optimal duration determined as 7 minutes. In this case the peak intensity in IR spectra is increased in 2–4 times (Table 21).

**Table 20. The peaks intensity change in dependence of Cu-C mesocomposite concentration**

| N | ν, см$^{-1}$ | $I_1/I_0$ | $I_{0,01}/I_0$ | $I_{0,001}/I_0$ | Отнесение |
|---|---|---|---|---|---|
| 1 | 1050 | 1,235 | 1,411 | 1,686 | C-O-C st |
| 2 | 1450 | 1,179 | 1,590 | 1,744 | C-H |
| 3 | 1776 | 1,458 | 1,347 | 1,691 | C=O st as |
| 4 | 1844 | 1,463 | 1,412 | 1,678 | C=O st sy |
| 5 | 2860-3090 | 1,182 | 1,545 | 1,750 | C-H |

**Table 21. The peaks intensity changes in IR spectrum of Cu-C mesocomposite depending on the duration of ultrasound treatment**

| ν (см$^{-1}$) | $I_7/I_0$ | $I_{10}/I_0$ | Отнесение |
|---|---|---|---|
| 1776,6 | 3,7932 | 0,7574 | C=O st as |
| 1844,1 | 2,5065 | 0,9115 | C=O st sy |
| 3039,1 | 2,3849 | 0,9589 | C-H st |

The charge (electron) quantization should lead to the macromolecule electron structure change and, as corollary, to change sub molecular structures of polymeric substances. Therefore the special film of nanostructured materials on the base of functional polymers such as polyvinyl alcohol, poly methyl metacrylate, polycarbonate, which contain metal carbon mesocomposite in the minute quantities ($10^{-1}$-$10^{-5}$%) are prepared. The films obtained are investigated by x-ray photoelectron spectroscopy (x-ray PES) and by atomic force microscopy (AFM). The investigations by x-ray PES show that the films based on polycarbonate have more changes of electron structure at the minute quantities introduction of Copper Carbon mesocomposite in comparison with other polymeric films because these films are more polarized. Below the C1s spectra for Cu/C nanocomposite and for nanostructured polycarbonate are presented (Figure 23).

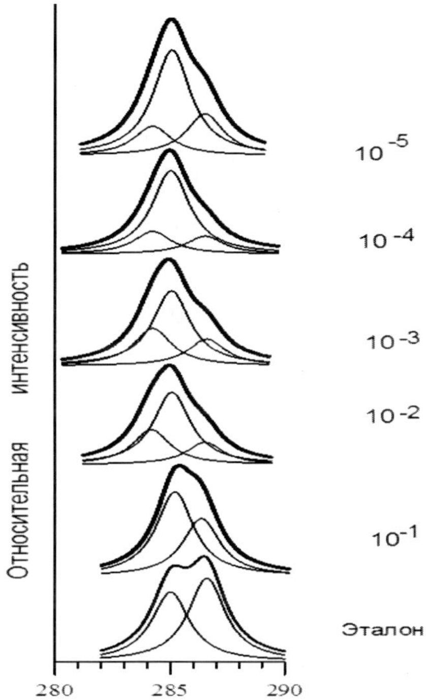


Figure 23. X-ray photoelectron C1s spectra of nanostructured polycarbonate modified by minute quantities ($10^{-1}$-$10^{-5}$%) of Cu-C mesocomposite.

The expansion of C1s spectra for nanostructured polycarbonate is possible owing to the determination of the energetic states for sp, $sp^2$ and $sp^3$ satellites. According to the results of C1s spectra for polycarbonate, contained the different minute quantities of Cu-C mesocomposite, can note that after concentration equaled to $10^{-2}$% of Cu/C mesocomposite the peaks correspondent to $sp^2$ and $sp^3$ peaks are appeared in these spectra. In other words, the "stamp" of mesocomposite which is used during modification is appeared. That "stamp" is observed also at the mesocomposite containing, equaled to $10^{-5}$%, in polycarbonate film. It's noted, that the relation between $sp^2$ and $sp^3$ peaks changes. For instance, the intensity of $sp^2$ hybridization carbon peak is upper the intensity of $sp^3$ peak in the concentration interval from 0,01 to 0,001% of mesocomposite. The change of concentration to $10^{-4}$% bring the proximity of intensities $sp^2$ and $sp^3$ peaks.

For the decision of question about the nanostructure influence on sub molecular composition structures the atomic force microscopy method is applied. Below some images of polycarbonate nanostructured films surface are presented. Polycarbonate is modified by Cu-C mesocomposite minute quantities (from $10^{-1}$ to $10^{-4}$%) (Figure 24).

As follow from AFM images (Figure 24), the surface layers structure strongly changes at the concentration of Cu-C MC equaled to $10^{-4}$ % the transition from two-dimensional level to three-dimensional level of sub-molecular structures orientation. This fact is confirmed by the growth of $sp^3$in comparison with $sp^2$ peak from x-ray photoelectron C1s spectra. It's interesting to observe the direction of carbon fibers in comparison with the direction of sub molecular structures orientation in the nanostructured polycarbonate surface layers.

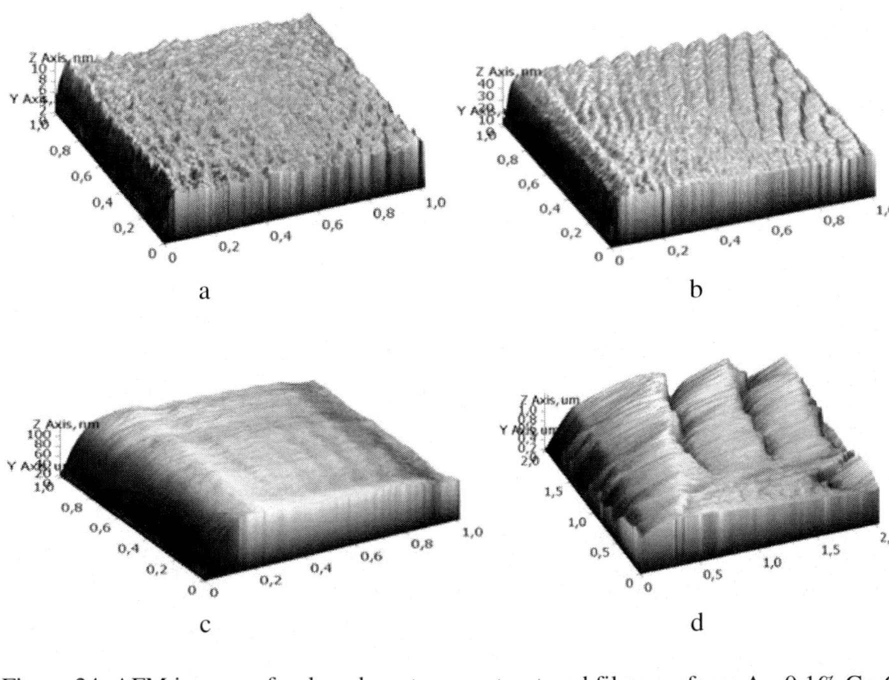


Figure 24. AFM images of polycarbonate nanostructured films surface: A - 0,1% Cu-C MC; B - 0,01% Cu-C MC; C - 0,001% Cu-C MC; D - 0,0001% Cu-C MC.

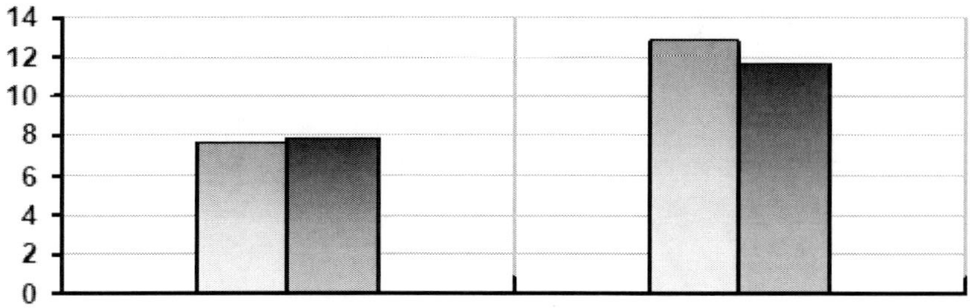


Figure 25. Comparison of adhesion strength for materials 1 (green) and 2 (dark blue) before (A) and after (B) the modification by Copper Carbon mesocomposites.

Thus, it's possible that the wave which initiates self organization process in polymeric composition is expensed from these fibers associated with metal cluster. The last leads to the correspondent orientation of sub molecular structures in nanostructured composite surface layers.

The self organization mechanism for polymeric compositions modified by the Metal Carbon mesocomposite minute quantities is concluded in the conditions creation for composition polarization, which bring the great change of electron and sub molecular structures of materials. Certainly, these changes influence on the modified materials properties.

Below the example of results for that modification with using of above considered suspensions will be presented. In this example, the epoxy compositions with different additives including mesocomposites were investigated. As cross-linking agent is used the fine dispersed suspension on based of isomethyl tetra hydro phthalates anhydrate and Copper Carbon mesocomposite.

On the Figure 25 the results of modification epoxy compounds (materials 1 and 2) are given.

The modification by Copper Carbon mesocomposite in quantity equaled to 0,005% improves the adhesion characteristics for the material 1 (green) on 59,77% and for the material 2 (dark blue) on 47,17%.

It's possible also other applications that nanostructures owing to its uncial structure and properties.

Below the results of some working-outs [40, 41] show:

1. The introduction of Metal Carbon nanostructures (0.005%) in the form of fine suspension into polyethylene polyamine or the mixture of amines into epoxy compositions allows increasing the thermal stability of the compositions by 75–100 degrees and consequently increase the application range of the existing products. This modification contributes to the increase in adhesive and cohesive characteristics of glues, lacquers and binders.
2. Hot vulcanization glue was modified with copper carbon and nickel carbon nanostructures with using Toluene as base for fine suspensions. On the test results of samples of four different schemes the tear strength $\sigma_t$ increased up to 50% and shear strength $\tau_s$ – up to 80%, concentration of metal carbon mesocomposite introduced was 0.0001-0.0003%.
3. Fine suspension of nanostructures was produced in dichloromethane and dichlorethane solutions to modify polycarbonate-based compositions. The introduction of 0.01% of Copper Carbon mesocomposites leads to the decrease in temperature conductivity of the material (in 1.5 times). The increase in the transmission of visible light in the range 400–500 nm and decrease in the transmission in the range 560–760 nm were observed.
4. When modifying polyvinyl chloride film by the fine suspension containing Iron Carbon mesocomposite (MC), the increase of the crystalline phase in the material was observed. The modified PVC film containing 0.0008% MC does not accumulate the electrostatic charge on its surface at the decreasing of electrostatic quantity more than two times. The material obtained completely satisfies the requirements applied to PVC films for stretch ceilings.
5. The introduction of Nickel Carbon mesocomposite (0.01% of the mass of polymer filled on 65% of silver microscopic particles) into the epoxy polymer hardened with polyethylene polyamine leads to the decrease in electric resistance to $10^{-5}$ Ohm·cm ($10^{-4}$ Ohm·cm without mesocomposite).

# 6. METAL CARBON MESOCOMPOSITES AS CORROSION INHIBITORS AND FIRE RETARDANTS

## 6.1. The Application Possibilities of Metal Carbon Mesocomposites as Inhibitors of Corrosion

From the foregoing follows that the metal carbon mesoscopic composites radiate the negative charged quants within media and stimulate the media polarization. This mesocomposites energetic action on media leads to media self organization and the density growth, and also chemical bonds formation. It's possible the high surface energy of metal carbon mesocomposites influences on the corrosion inhibitor reactivity and the adsorbed layer protective properties.

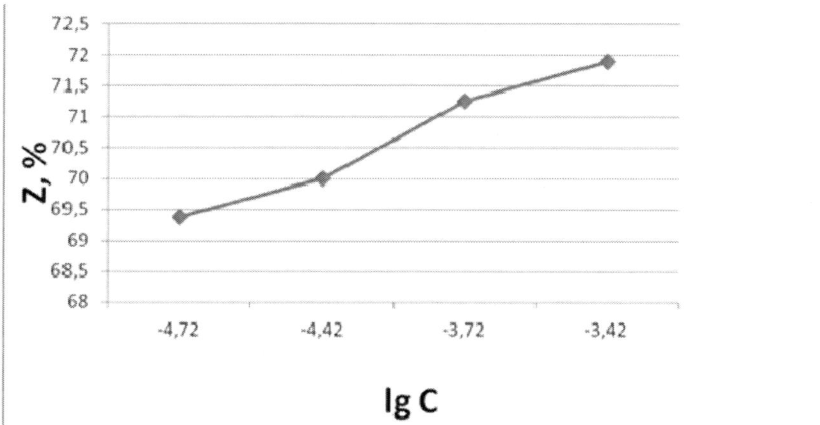


Figure 26. The dependence of the inhibitor protection degree at steel corrosion from Copper Carbon mesocomposite concentration logarithm.

According to [38] the addition of Copper Carbon mesocomposite to such corrosion inhibitor as 1-morpholine methyl cyclohexyl amine is effective at the quantity equaled to 0,001 mg/m$^3$. In this paper it's shown that the joint using of Copper Carbon mesocomposite with previous corrosion inhibitor improves the protective properties in different corrosive media. The mesocomposite content increasing in the mixture with corrosion inhibitor promotes to the growth of protection degree at the corrosion.

The dependence of the inhibitor protection degree from Copper Carbon mesocomposite concentration logarithm presents (Figure 26).

The experimental results show that the increasing of concentration of mesocomposite (Cu-C MC) leads to the growth of inhibitor protection efficacy.

The mesocomposite containing inhibitor efficacy relations in comparison with initial inhibitor are increased. These relations changes are shown as follows:

| **Inhibitor content:** | **25** | **15** | **10** | **5** |
|---|---|---|---|---|
| Relations of combined (with Cu-C MC) inhibitor efficacy in comparison with the initial inhibitor | 1,06 | 1,33 | 1,5 | 5,4 |
| Protection degree: | | | | |
| Inhibitor, Z% | 80 | 60 | 50 | 5 |
| Inhibitor with 0,001mg/dm3 of Cu-C MC, Z% | 85 | 80 | 75 | 27 |

These changes are also shown on the Figure 27.

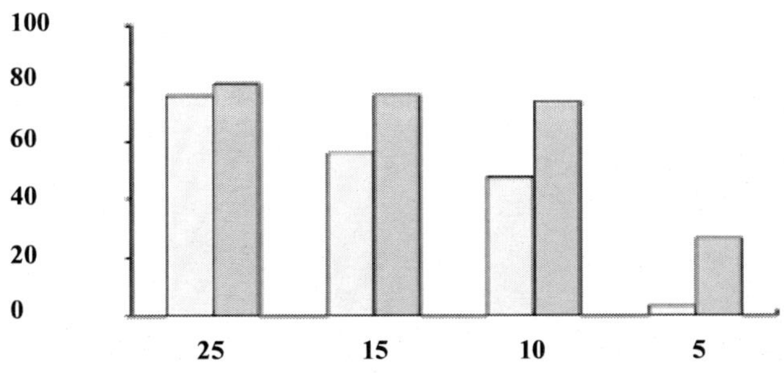


The inhibitor concentration, mg/m$^3$.

1 - without mesocomposite; 2 – with mesocomposite (0,001 mg/m$^3$).

Figure 27. The dependence of the inhibitor protection degree from concentrations of inhibitor and the mixture of inhibitor and mesocomposite.

Therefore the following conclusion can be made: Cu-C mesocomposite can be effective inhibitor of corrosion. Mesocomposite fulfills two functions: 1) for corrosion active agent decrease; 2) for the protective film creation on the material surface. These functions are carried owing to the presence on the mesocomposite carbon shell the unpaired electrons and double bonds. Actually the stable coating with the protection degree approximately 95% can be obtained from the oil fine suspension of the Phosphorus containing mesocomposite at the multi graded thermal processing (with the temperature increasing to 500°C).

## 6.2. The Fire Retardants or Synergists to Fire Retardants on the Base pf Metal Carbon Mesocomposites

### *6.2.1. Fire Resistant Glues Based on Phenol-Formaldehyde Resins*

In this paragraph the results of modification of fireproof materials, fire resistant intumescent coatings and glues, modified by nanostructures are considered. In the beginning the example of application of Metal Carbon mesocomposite for the obtaining of fire resistant materials is brought. In this case the modification of phenol-formaldehyde glues for the obtaining from them intumescent fire resistant glues is done.

The glues based on phenol-formaldehyde resins (BF–19) are modified by Copper Carbon mesocomposite and also by Phosphorus containing analog. It's noted that the introduction of these mesocomposites into the glue significantly decreases the material flammability. The samples with Phosphorus containing mesocomposites have better test results. In the case, when Phosphorus containing mesocomposite is introduced into the glue, foam coke is formed on the sample surface during the fire exposure. The coating flaking off after flame exposure is not observed at the coating preserved good adhesive properties even after the flammability test.

The Phosphorus presence on the mesocomposite surface allows improving the mesocomposite structure and increases the activity in different liquid media thus increasing their influence on the material modified. The modification of coatings by mesocomposites obtained finally results in improving their fire-resistance and physical and chemical characteristics.

The modification of BF-19 glue by Metal carbon mesocomposites and Phosphorus containing analogs is considered below.

The glue BF-19 is intended for gluing metals, ceramics, glass, wood and fabric in hot condition, as well as for assembly gluing of cardboard, plastics, leather and fabrics in cold condition. The glue composition: organic solvent, synthetic resin (phenol-formaldehyde resins of new lacquer type), synthetic rubber.

When modifying the glue composition, at the first stage the mixture of alcohol suspension (ethyl alcohol + Cu C MC $_{mod}$) and Ammonium Polyphosphate (APPh) is prepared. At the same time, the mixtures containing ethyl alcohol, Cu-C MC and APPh, ethyl alcohol and APPh are prepared. At the second stage the glue composition is modified by the introduction of phosphorus containing compositions prepared into the glue BF-19.

For flammability determination the following method of preparation samples is used:

> The samples to be tested are the plates with the dimensions 150×15×3 mm. The plates consist of foam polyethylene and paper glued together with phosphorus containing glue modified with Metal Carbon mesocomposites with and without phosphorus. At the same time, check samples are prepared. These are the plates of foam polyethylene and paper glued together with the glue BF-19 filled with ammonium polyphosphate (APPh) with phosphorus content in the glue 3, 4, 5% from its mass.

For the testing of sample flammability the estimation method on the determination of the lengths of carbonized parts with 1.5 minute flame exposure is used. To compare the results of coating flammability, three samples of glue are selected and tested.

The test results revealed that the length of carbonized part of the samples containing APPh (Ammonium Polyphosphate) and exposed to burner flame for 1.5 minutes can be about 8.5 cm with 3% phosphorus content in the sample (Table 22).

**Table 22. Results of testing samples containing Ammonium Polyphosphate**

| Sample No | Sample composition | Phosphorus content, % | Flame exposure time, min | Length of carbonized part of samples, mm |
|---|---|---|---|---|
| 1 | APPh | 5 | 1.5 | 65.33 |
| 2 | APPh | 4 | 1.5 | 82 |
| 3 | APPh | 3 | 1.5 | 84.67 |
| | Average | | | 77.33 |

**Table 23. Results of testing samples of modified compositions with mesocomposite**

| Sample No | Sample composition | Phosphorus content, % | MC content, % | Flame exposure time, min | Length of carbonized part of samples, mm |
|---|---|---|---|---|---|
| 4 | Cu/C NC pure + APPh | 5 | 0.00025 | 1.5 | 15.33 |
| 5 | Cu/C NC pure + APPh | 4 | 0.0002 | 1.5 | 23.43 |
| 6 | Cu/C NC pure + APPh | 3 | 0.00015 | 1.5 | 26.66 |
| | Average | | | | 21.81 |

The composition of sample coating: glue BF-19 + APPh + Cu/C MC $_{mod}$.

The tests of check samples confirmed that with the phosphorus content increase in composition the length of carbonized parts of samples goes down.

*The composition of sample coating: glue BF-19 + APPh + Cu/C MC* $_{pure}$.

The next step was to test samples containing mesocomposites with the decreasing of phosphorus content. The average value of the carbonized part of the samples was 21.81 mm. The test results (Table 23) allow making the conclusion that nanocomposite inclusion significantly decreases the material flammability (in 3.5 times).

Phosphorus containing samples of Cu/C MC had better flammability test results than samples with APPh and samples containing APPh and Cu/C $MC_{pure}$. The length of the carbonized part of the samples was less by 3 mm in the average [39]. The average value of the carbonized part of the samples was 18.89 mm (Table 24). Thus it can be concluded that the inclusion of Phosphorus in mesocomposite decreases the material flammability to a greater extent than mesocomposite which do not contain Phosphorus.

**Table 24. Results of testing samples of compositions modified by Phosphorus containing mesocomposites**

| Sample No | Sample composition | Phosphorus content, % | NC content, % | Flame exposure time, min | Length of carbonized part of samples, mm |
|---|---|---|---|---|---|
| 7 | Cu/C NC $_{mod}$ + APPh | 5 | 0.00025 | 1.5 | 14.67 |
| 8 | Cu/C NC $_{mod}$ + APPh | 4 | 0.0002 | 1.5 | 16.67 |
| 9 | Cu/C NC $_{mod}$ + APPh | 3 | 0.00015 | 1.5 | 25.33 |
| | Average | | | | 18.89 |

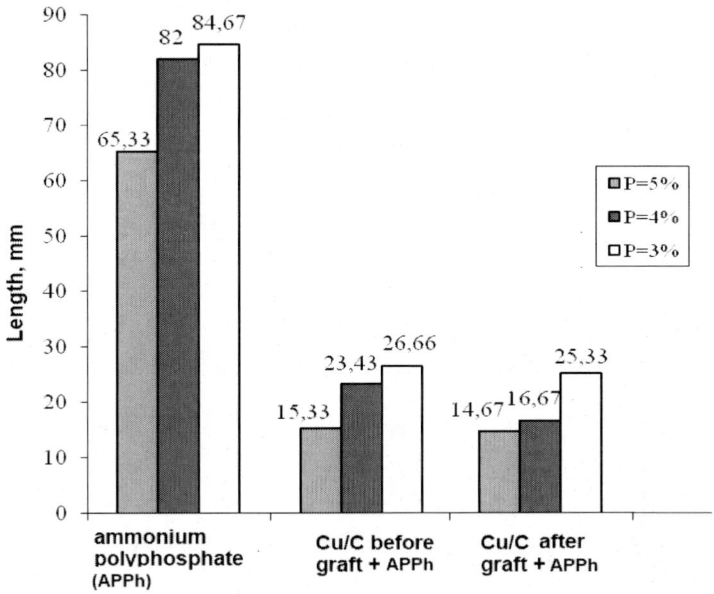


Figure 28. Diagram of the lengths of carbonized parts of the samples depending on phosphorus content in the composition.

From the data demonstrated based on the test results it can be concluded that mesocomposite inclusion into the glue composition significantly decreases the material flammability. The length of the carbonized part of the samples modified with mesocomposites was in 4.1 times in the average less in comparison with similar parameters of the samples not containing mesocomposites. The samples with phosphoric groups in mesocomposites have better test results (Figure 28).

The coating flaking off after flame exposure is not observed, i.e., the coating preserves good adhesive properties even after the flammability test.

When the intumescent glue composition is modified with nanostructures, the material is structured with the formation of crystalline regions. In turn, such structuring under the influence of mesoscopic systems results in the increased physical and mechanical characteristics, including their stability against high and low temperatures.

### *6.2.2. Properties Studies of Fire Proof Coating Based on Epoxy and Melamine Formaldehyde Resines*

The obtaining and investigations of properties for the intumescent fire proof materials based on epoxy and melamine formaldehyde resins are considered. In both cases the Phosphorus containing Copper Carbon mesocomposite together with Ammonium Polyphosphate (P → Cu-C MC) is used. At the production of fire proof materials based on the epoxy resins the mesocomposite P → Cu-C MC is introduced into polyethylene polyamine for the fine dispersion preparation. In the case when the fire proof coating is based on the melamine formaldehyde resin the P → Cu-C MC is introduced into melamine formaldehyde lacquer. For the testing of correspondent coatings such metal substances as steel and aluminum are applied. It's necessary to note that the last coating will be simultaneously fire proof and anticorrosive coating. This is especially significant for the protection of aluminum. Both coating are studied on the adhesive durability and fire proof efficiency. These tests are necessary because it's known that the coating adhesive durability is decreased at the fire action on coating. However when the heat capacity of coating is increased at the influence of mesocomposite the coating adhesion strength is conserved. At the investigations of correspondent characteristics of coatings the following methods are applied: for the tests on the fire proof efficiency – ASTM E–14–10 method, and for the tests on the adhesive durability – GOST 14753–73 method (Russian).

The studies on the ASTM E–14–10 method are carried out on samples which have plate forms with the following sizes: length – 50-55 mm, width – 10-15 mm, thickness - 1-1,3 mm. The fire resistant composition covers the correspondent plate. The coating layer thickness on plate compiles 1 mm. The samples are cross linked at normal conditions during 24 h. In the case when fire resistant composition is presented as epoxy resins modified by Ammonium Polyphosphate (APPh) with mesocomposites or without them the flame of gas burner is used at these tests. The characteristics of fire resistant protection can be height of foam coke formed from above fire resistant compositions. The height of foam coke is measured from the surface of steel plate to upper level of foam coke layer. The results of the studies of fire proof efficiency for the coatings containing APPh in the comparison with the coatings modified by the Copper Carbon mesocomposite (Cu-C MC) or by the Phosphorus containing Cu-C MC (P → Cu-C MC) are presented below (Figure 29).

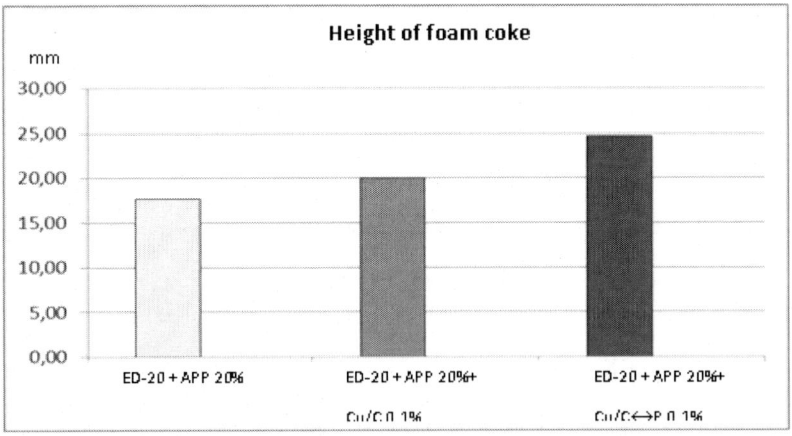


Figure 29. The dependence of the samples foam cokes height from the fire proof coating content.

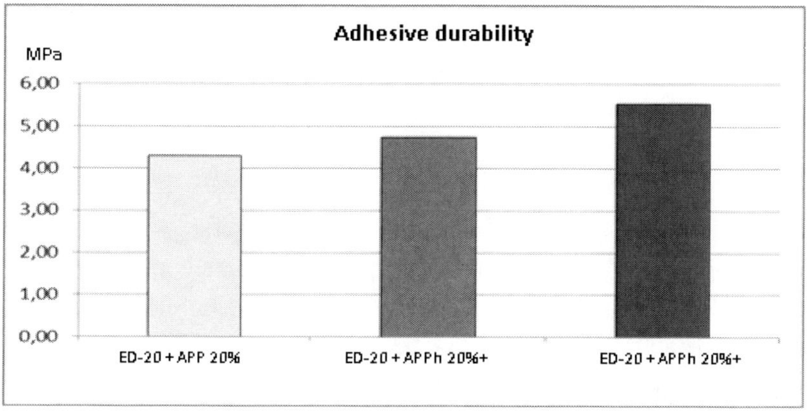


Figure 30. The dependence of samples adhesive durability from contents of them.

In correspondence with results obtained, the height of foam cokes samples which are modified by means of phosphorus containing Copper Carbon mesocomposite (c, red color), is upper, approximately, on 47% than initial sample (a, yellow color).

For the determination of Copper Carbon mesocomposites introduction efficiency the following compounds are prepared:

1. ED – 20 – APPh (20%) with polyethylene polyamine – initial sample
2. ED – 20 – APPh (20%) with fine dispersed suspension of Cu-C MC
3. ED – 20 – APPh (20%) with fine dispersed suspension of P containing Cu-C MC.

The quantity of mesocomposites which are introduced in compositions is equaled to 0,1%. This quantity of mesocomposites is considered as maximum for the getting of positive effect without negative consequences. The diagram compounds adhesive durability dependence from the content and forms is presented on Figure 30.

According to Figure 30 the introduction Phosphorus containing Copper Carbon mesocomposite in coating composition (sample c, red) leads to the increasing of adhesive durability on 30% in the comparison with the fire proof coatings without the presence of mesocomposite (sample a, yellow).

Thus, on the base of above experiments, more effective fire proof coating is intumescent coating based on epoxy resins with APPh modified by phosphorus containing Copper Carbon mesocomposite that testify about the increasing of Copper Carbon mesocomposite activity after modification of initial mesocomposite by ammonium polyphosphate. The application of these active nanosystems for the obtaining of fire proof intumescent coatings containing epoxy resins and ammonium polyphosphate is perspective.

At the investigations of coatings based on melamine formaldehyde lacquer modified by Ammonium Polyphosphate (APPh) with Copper Carbon mesocomposite (Cu-C MC) the following optimal quantities of modifiers were established: APPh – 15%; Cu-C MC – 0,008%. In this case the dense protective layer formation on steel is determined by means of Atomic Force Microscopy (AFM) method.

The coating fire proof efficiency properties improvement is observed:

- The mass losses decreasing on 43% at the burning.
- The burnt part length abridgement on 66% in the burning conditions.
- The burnt velocity decreasing on 45% in the same conditions.

Thus the coating based on melamine formaldehyde lacquer modified by 15% of Ammonium Polyphosphate with 0,008% of Copper Carbon mesoscopic composite can be applied as intumescent fire proof and anticorrosive coating for metals including steel and aluminum.

## CONCLUSION

The presented review is dedicated by the consideration of properties and the possible applications of new class of such mesoscopic particles as Metal Carbon mesoscopic composites and their modified analogs. This review consists from row paragraphs in which the consecutive description of structures and properties for Metal Carbon mesocomposites and their modified analogs is done. The paragraphs, beginning with the second, are accompanied the correspondent examples of the described mesocomposites applications in the different spheres. In the first paragraph the structures and properties for initial Metal Carbon mesocomposites are considered. The unique structures of Metal Carbon mesocomposites are explained by reactivity of Carbon shells which contain the poly acetylene and carbine fragments with unpaired electrons on the joints of fragments. The stability of shells is possible because of the interactions of double bonds, which take place in fragments, with metal clusters within mesocomposites. The increasing of Carbon shell activity is obtained after the reduction oxidation reactions at the modification of initial Metal Carbon mesoscopic composites by reagents containing the positive charged atoms (or oxidizers). These processes as it's established in experiments lead to the shift of electrons on the more high energetic levels and the formation of unpaired electrons that is accompanied by the growth of atomic magnetic moments of mesocomposite cluster metals. These phenomena of structures and energetic characteristics of mesocomposites obtained cause their possibilities for applications in the different fields.

The examples of following applications in radical, red ox and addition processes as catalysts, reagents and also inhibitors as well as additives and modifiers improving properties of materials (inorganic and organic polymeric materials), adhesives and glues, fire proof systems, corrosion inhibitors, medicine magnetic transport remedies, stimulators of plant growth are presented. The Metal Carbon mesoscopic composites owing to their magnetic characteristics can be used in the electromagnetic radiation focal systems. This unique scientific trend discover new era in the development of novel nanostructures application widening.

## REFERENCES

[1] Kodolov V. I., Khokhrikov N. V. (2009) *Chemical physics of the processes of formation and transformation of nanostrustures and*

*nanosystems (2 volumes: Vol. 1, p. 360. Vol. 2, p. 415)*. Izhevsk: Izhevsk State Agricultural Academy.

[2] Shabanova I. N., Kodolov V. I., Terebova N. S., Trineeva V. V. (2012) *X ray electro spectroscopy in investigation of metal/carbon nanosystems and nanostructured materials*. (P. 252) Izhevsk–Moscow: Publ. "Udmurt University."

[3] Kodolov V. I., Trineeva V. V. (2013) Fundamental definitions for domain of nanostrustures and metal/carbon nanocomposites in Book: *Nanostructure, nanosystems and nanostructured materials. Theory, production and development* (Pp. 1–42) Toronto, Canada – New Jersey, USA: Apple Academic Press.

[4] Khokhriakov N. V., Kodolov V. I., Korablev G. A. et al. (2013) *Prognostic investigations of metal or carbon nanocomposites and nanostructures synthesis processes characterization*. Ibid. (Pp. 43–99).

[5] Kodolov V. I., Trineeva V. V., Blagodatskhikh I. I. et al. (2013) *The nanostructures obtaining and the synthesis of metal or carbon nanocomposites in nanoreactors*. Ibid. (Pp. 101–146).

[6] Kodolov V. I., Akhmetshina L. F., Chashkin M. A. et al. (2013) *The functionalization of metal or carbon nanocomposites or the introduction of functional groups in metal/carbon nanocomposites.* Ibid. (Pp. 147–175).

[7] Shabanova I. N., Terebova N. S. Kodolov V. I. et al. (2013) *The investigation of metal or carbon nanocomposites electron structure by X ray photoelectron spectroscopy*. Ibid. (Pp. 177–230).

[8] Kodolov V. I., Khokhriakov N. V., Trineeva V. V. et al. (2013) *Computation modeling of nanocomposites action on the different media and on the composition modification processes by metal/carbon nanocomposites*. Ibid. (Pp. 231–286).

[9] Kodolov V. I., Lipanov A. M., Trineeva V. V. et al. (2013) *The changes of properties of materials modified by metal/carbon nanocomposites*. Ibid. (Pp. 327 – 373).

[10] Kodolov V. I., Trineeva V. V. (2013) Theory of modification of polymeric materials by super small quantities of metal/carbon

nanocomposites.// *Chemical Physics & Mesoscopy, V.* 15. N 3. Pp. 351–363.

[11] Kodolov V. I., Trineeva V. V. (2012). Perspectives of idea development about nanosystems self organization in polymeric matrixes In book "*The problems of nanochemistry for the creation of new materials*" - Torun, Poland: IEPMD. (Pp. 75-100).

[12] Akhmetshina L. F., Lebedeva G. A., Kodolov V. I. (2012). Phosphorus containing metal/carbon nanocomposites and their application for the modification of intumescent fireproof coatings. // *Journal of Characterization and Development of Novel Materials, 2012. V. 4. N 4.* Pp. 451–468.

[13] Kodolov V. I., Kovyazina O. A., Trineeva V. V., Vasilchenko Yu. M., Vakhrushina M. A., Chmutin I. A. On the production of metal/carbon nanocomposites, water and organic suspensions on their basis.// VII International Scientific-Technical Conference "Nanotechnologies to the production – 2010."/ *Proceedings. Fryazino*, 2010. Pp. 52-53.

[14] Chashkin M. A. Peculiarities of modification by metal/carbon nanocomposites for cold hardened epoxy compositions and the investigation of properties of polymeric compositions obtained – *Thesis of cand. diss.* – Perm: PNSPU, 2012. 17p.

[15] *Patent 2337062 Russia Technique of obtaining carbon nanostructures from organic compounds and metal containing substances* / Kodolov V. I., Kodolova V. V. (Trineeva), Semakina N. V., Yakovlev G. I., Volkova E. G., et al. declared on 28.08.2006, published on 27.10.2008.

[16] Kodolov V. I., Trineeva V. V., Kovyazina O. A., Vasilchenko Yu. M. Production and application of metal/carbon nanocomposites.// In book "*The problems of nanochemistry for the creation of new materials*" – Torun, Poland: IEPMD, 2012. Pp. 23 – 36.

[17] Akhmetshina L. F., Kodolov V. I., Tereshkin I. P., Korotin A. I. The influence of carbon metal containing nanostructures on strength properties of concrete composites. // *Internet Journal "Nanotechnologies in Construction,"* 2010. N 6. Pp. 35–46.

[18] Kodolov V. I., Khokhriakov N. V., Trineeva V. V., Blagodatskikh I. I. (2008) Activity of nanostructures and its display in nanoreactors of

polymeric matrixes and in active media. // *Chem. Phys. & Mesoscopy, V. 10. N 4*. Pp. 448–460.

[19] Kodolov V. I., Khokhriakov N. V., Trineeva V. V., Blagodatskikh I. I. (2010) Problems of nanostructures activity estimation, nanostructures directed production and application. – in *Nanomterials yearbook – 2009. From nanostructures, nanomaterials and nanotechnologies to nanoindustry* – N. Y.: Nova Science Publ., Inc. Pp. 1–18.

[20] Kodolov V. I., Blagodatskikh I. I., Lyakhovitch A. M. et al. (2007). Investigation of the formation processes of metal containing carbon nanostructures in nanoreactors of polyvinyl alcohol at early stages // *Chem. Phys. & Mesoscopy, V. 9. N 4*. Pp. 422–429.

[21] Trineeva V. V., Lyakhovitch A. M., Kodolov V. I. (2009) Forecasting of the formation processes of carbon metal containing nanostructures using the method of atomic force microscopy. // *Nanotechnics. Iss. 4(20)*. Pp. 87 – 90.

[22] Khokhriakov N. V., Kodolov V. I. Quantum-chemical modeling of nanostructure formation // *Nanotechnics, 2005*. № 2. P. 108–112.

[23] Kodolov V. I., Khokhriakov N. V., Nikolaeva O. A., Volkov V. L. Quantum-chemical investigation of alcohols dehydration and dehydrogenization possibility in interface layers of vanadium oxide systems // *Chemical physics and mesoscopy, 2001. V. 3. № 1. P. 53–65.*

[24] Kodolov V. I., Didik A. A., Volkov A. Yu., Volkova E. G. Low-temperature synthesis of copper nanoparticles in carbon shell // *Bulletin of HEIs. Chemistry and chemical engineering, 2004. V. 47. № 1. P. 27–30.*

[25] Lipanov A. M., Kodolov V. I., Khokhriakov N. V. et al. Challenges in the production of nanoreactors for the synthesis of metallic nanoparticles in carbon shells // *Alternative energetic and ecology (ISJAEE), 2005. № 2(22)*. P. 58–63.

[26] Nikolaeva O. A., Kodolov V. I., Zakharova G. S. et al. Method of obtaining carbon-metal-containing nanostructures. / *Patent of the RF 2225835*, registered on 20.03.2004.

[27] Didik A. A., Kodolov V. I., Volkov A. Yu. et al. (2003) // *Inorganic Materials, 39(6).* Pp. 693 – 697.

[28] Kodolov V. I., Trineeva V. V. The metal/carbon nanocomposites influence mechanisms on media and compositions. – In Book "*Nanostructures, nanomaterials and nanotechnologies to nanoindustry*" – Toronto, New Jersey: Apple Academic Press, 2015, Pp. 171– 185.

[29] Shabanova I. N., Kodolov V. I., Terebova N. S. et al. x-Ray photoelectron study of the influence of the amount of carbon nickel containing nanostructures on the degree of the poly methyl metacrylate modification. – In Book "*Multifunctional materials and modeling*" – Canada-USA: AAP, 2016, Pp. 211–218.

[30] Kodolov V. I., Trineeva V. V. Self organization in processes under action of super small quantities of metal/carbon nanocomposites. Review on investigation results. – In Book "*Multifunctional materials and modeling*" – Canada-USA: AAP, 2016. – 343p. – Pp. 263–331.

[31] Kodolov V. I., Trineeva V. V., Pershin Yu.V. et al. *Method of metal carbon nanocomposites obtaining from metal oxides and polyvinyl alcohol.* –Pat. RU 2018122 001, 2020.

[32] Mustakimov R. V., Kodolov V. I., Shabanova I. N., Terebova N. S. Modification of copper carbon nanocomposites with the using of Ammonium Polyphosphate for the application as modifiers of epoxy resins.// *Chemical Physics & Mesoscopics*, 2017. V. 19. N 1. Pp. 50–57.

[33] Kodolov V. I., Trineeva V. V., Kopylova A. A. et al. Mechanochemical modification of metal carbon nanocomposites. // *Chemical Physics & Mesoscopics*, 2017. V. 19. N 4. Pp. 569–580.

[34] Kodolov V. I., Trineeva V. V., Terebova N. S. et al. The change of electron structure and magnetic characteristics of modified copper carbon na nocomposites.// *Chemical Physics & Mesoscopics*, 2018. V. 20. N 1. Pp. 72–79.

[35] Merzlyakova V. M., Lapin A. A., Kodolov V. I. Efficiency of metal carbon nanocomposite application in lily growing under protected soil conditions. – In Book "*Nanoscience and Nanoengineering. Novel*

*applications.* – Toronto-New Jersey: Apple Academic Press, 2019 – 355p. – Pp. 177–188.

[36] Karavaeva N. M., Pershin Yu. V., Kodolov V. I. et al. Change of morphology and swelling of cured epoxy compositions upon their modification with minute quantities. // *Pol. Sci., ser. D.*, 2019. V. 12. N 2. Pp. 179 – 181.

[37] Kodolov V. I. et al. Modified magnetic metal carbon mesoscopic composites with bio active substances. // *Chemical Physics & Mesoscopics*, 2019. V. 21. N 3. Pp. 446– 454.

[38] Pletnev M. A., Ovechkina O. A., Buldakova N. S. et al. The influence of metal carbon nanocomposites on protective action of corrosion inhibitors. // *Intellectual systems in engineering*, 2014. N 1(23). Pp. 150–152.

[39] Mustakimov R. V. *The method of modification of metal carbon nanostructures by ammonium poly phosphate*. Pat. N 2694092, 2019.

[40] Kodolov V. I., Semakina N. V., Trineeva V. V. *Introduction in science about nanomaterials. – Monograph.* – Izhevsk: Publisher – M. T. Kalashnikov Izhevsk State Technical University, 2018 – 476p.

[41] Kodolov V. I., Kodolova –Chukhontseva V. V. *Fundamentals of Chemical Mesoscopics. – Monograph.* - Izhevsk: Publisher – M. T. Kalashnikov Izhevsk State Technical University, 2019 – 218p.

[42] Pat. RU N 2490027, 2017.

[43] Wang J. Q., Wu W. M., Feng D. M. *The introduction of Electron Spectroscopy (XPS/XAES/UPS).* – Beijing: National Definite Industry Press, 1992, 640p.

In: Innovations and Challenges in Modern … ISBN: 978-1-53618-181-4
Editor: A. K. Haghi 

*Chapter 4*

# ANTIBIOTIC SENSITIVITY OF *STAPHYLOCOCCUS AUREUS* FROM *PROSOPIS JULIFLORA*, *CASSIA OCCIDENTALIS* AND *TEPHROSIA PURPUREA*

***Aishwarya A. Andhare and Ravindra S. Shinde****
Department Of Microbiology, Biotechnology and Chemistry,
Dayanand Science College, Maharashtra, India

## ABSTRACT

*Staphylococcus aureus* is resistance commonly used antibiotics and is being added to their ability to acquire and disseminate antimicrobial resistant property determinants in nature. This study was carried out to determine the antibiotic sensitivity of *Staphylococcus aureus* which was isolated from milk samples of mastitic and healthy cattle from dairy. Isolation, Identification and Characterization of *Staphylococcus aureus* was done by Marphological and Biochemical characters. The primary purpose of this study was to determine Antibiotic Sensitivity of *Staphylococcus aureus* by using extracts of *Prosopis juliflora*, *Cassia occidentalis* and *Tephrosia purpurea.* Antimicrobial properties of plant

* Corresponding Author's Email: rshinde.33381@gmail.com.

extracts were analyzed by determining Zone of Inhibition (ZOI) and Minimum Inhibitory Concentration (MIC). Antibiogram pattern of the isolated *Staphylococcus aureus* was observed to be Susceptible, Intermediate and Slight resistant by *Prosopis juliflora*, *Tephrosia purpurea and Cassia occidentalis* respectively. From the results, it has been indicated that the extracts of *Tephrosia purpurea*, *Prosopis juliflora* and *Cassia occidentalis* are having effectiveness to use in the management of borne diseases. Further phytochemical analysis is required to identify the active components of plant extracts showing antimicrobial activity.

**Keywards:** zone of inhibition, antimicrobial-resistant, antibiotic sensitivity, *staphylococcus aureus*, antibiogram pattern

## 1. INTRODUCTION

The definition of Food borne diseases made by World Health Organization (WHO) is "diseases of infectious or toxic nature caused by the consumption of food or water from non hygienic area". The pathogenesis of pathogrn causing food borne poisoning depends on their capacity to produce toxin chemicals after ingestion or intoxication. Among the bacteria which are predominantly taking part in induction of diseases, *Staphylococcus aureus* is one of the major agent of gastroenteritis resulting from the intake of unhygienic or contaminated food. The food poisoning by Staphylococcus is due to the absorption of entero-toxins secreted by *Staphylococcus* in the food [1].

*Staphylococcus aureus* are Gram-positive cocci (Size: 0.5-1.5 m diameter), which has polysaccharide capsule but some of the strains has no polysaccharide capsule. They does not contain flagella, non-spore forming and facultative anaerobes. They produces coagulase and catalase enzymes [2, 3]. From several decades, marine water is microbial contaminated and is predicted to be the reason for millions of acute respiratory and gastrointestinal diseases [4], also it causes various skin infections [5]. Despite, *S. aureus* is typically known to be opportunistic because it is commensal organism. *S. aureus* is a potentially harmful human pathogen associated with both nosocomial and community-acquired infections, and it has adapted a nature to be resistant to most antibiotics. Previous studies of *S. aureus* in marine environments have linked swimmers to the dissemination of *S. aureus* in marine water [8], via the climb out of the bacterium from their body organs such as, nose, skin, and respiratory tract [9]. On recreational beaches, *S. aureus* has occasionally been found in high abundance in both water and sand, which can be directly associated with bather density and human activities around the beach [9-11]. *Staphylococcus aureus* cause disease through the production of toxin or through direct invasion and destruction of tissue. Infections caused by *S. aureus* remain a significant cause of mortality and morbidity in tropical countries [12].

The principal site of staphylococcal colonization is the anterior nares. It has been observed that if repeated cultures are performed, up to 80% of adults are found to harbor *S. aureus* in the nose at one time or the other. However, in most persons, the carrier state is transient, but 20 to 40% of adults remain colonized for months or even years [13]. Increased nasal colonization rates have been noted in insulin dependent diabetes [14], individuals on haemodialysis [15], those on ambulatory peritoneal dialysis [16], intravenous drug users [17] and patients receiving routine allergy injections [18]. It has also been suggested that patients with symptomatic human immunodeficiency virus infection have an increased colonization risks [19]. Staphylococci have a record of developing resistance quickly and successfully to antibiotics. This defensive response is a consequence of the acquisition and transfer of antibiotic resistance plasmids and the possession of intrinsic resistance mechanisms [20]. The importance of *Staphylococcus aureus* as a persistent nosocomial and community acquired pathogen has become a global health concern. It has a remarkable capability of evolving different mechanisms of resistance to most antimicrobial agents [21].

*Staphylococcus aureus* is one of the major cause for life-threatening and most affecting human infections acquired within the community and in hospitals [23, 24]. The first methicillin-resistant *Staphylococcus aureus* (MRSA) infection was described in 1961 [25] and since then, human infections caused by multi-drug-resistant MRSA have become common [26]. The MRSA phenotype results from the acquisition of the staphylococcal cassette chromosome mec (SCC*mec*), which harbors the mecA gene that encodes the low-affinity penicillin-binding protein 2a (PBP2a) [27]. The structure of numerous SCCmec elements reveals that this element is following multiple evolutionary trajectories [for review see 28, 29].

Antimicrobial agents are frequently used in the treatment of bacterial infections; however, multiple drug resistance in human pathogens has been developed due to use of commercial antimicrobial drugs commonly used to cure such diseases [30]. Hence, discovery of new antibacterial substances either synthetic or natural with minimal side effects is necessary. Increasing popularity of plant-based traditional medication has led researchers to consider the natural compounds, of plant origin, as a valuable source for medicines [31]. Natural products of the plants have been used to treat various illnesses [32]. Plants contain a group of chemicals that protect them against various microbes and some of their extracts have been used against many types of infections. Even today plants are the almost exclusive source of drugs for the majority of people in the world [33-35]. *P. juliflora* has been used as a traditional treatment for catarrh, cold, diarrhea, dysentery, flu, hoarseness, inflammation, measles, sore throat, and hepatic and ocular problems [35, 37]. Wounds can be treated with the decoction obtained from leaf and seed extracts [35, 38]. The syrup, made from ground pods of *P. juliflora*, is used for nourishment of underweight kids or those suffering from retardation in motor development. The syrup may raise lactation as well [35]. Tea prepared from *P. juliflora* is believed to be useful in healing of digestive disorders and skin wounds. *P. juliflora* is a good source of piperidine alkaloids. Several alkaloids including juliflorine, julifloricine and juliflorididine, juliprosine, juliprosinene and juliflorinine, 3' oxojuliprosopine, sceojuliprosopinol, 3-oxojuliprosine and 3'-oxo-juliprosine have been isolated from various parts of *P. juliflora* and have shown to be pharmacologically active [35].

*Tephrosia purpurea* belongs to the family Fabaceae. English name is purple *Tephrosia*, vernacular name is Kolangi. It is a perennial herb found throughout the Indian subcontinent. The plant is 30-50cm height. Leaves are bipinnate, the flowers are pink and the effective parts were used as medicines are leaves, stems and roots. The aerial parts of *Tephrosia purpurea* were used to treat pathological conditions like hydrophobia, asthma, cough, heart, lung diseases, kidney problems, mouth ulcer and piles. *Tephrosia purpurea* is a medicinal plant, considered highly useful in bilious febrile attacks and obstruction of liver and spleen [38]. According to Ayurveda literature, this plant has the property of healing all types of wounds. It is an important component of some preparation such as Tephroli and Yakrifit used for liver disorders [39-41]. The *Tephrosia purpurea* leaves are reported to be useful in jaundice [42]. *Tephrosia purpurea* has been shown to possess antimicrobial activity [43], insecticidal and repellent activity [44], antilithiatic activity [45], antihyperglycemic and anti lipid peroxidative effect [46]. The plant was found to contain rutin, quercetin, lupeol, retinoid mainly degulin, elliptone, rotenone and tephrosin [47, 48].

*Cassia occidentalis* L belongs to family Caesalpiniaceae which is called as Kasmard in Sanskrit while Kasondi in Hindi and Coffee Senna in English. It is an ayurvedic plant with huge medicinal importance. Leaves of *C. occidentalis* plant have ethnomedicinal importance like paste of leaves is externally applied on healing wounds, sores, itch, cutaneous diseases, bone fracture, fever, ringworm, skin diseases and throat infection. Earlier researchers have carried out several pharmacological investigations and showed that *C. occidentalis* leaf extracts have antibacterial [49, 50], antimalarial [51], antimutagenic [52], antimutagenic [53], antiplasmodial [54] anticarcinogenic [55] and hepatoprotective [56] activity. However, studies of this plant showed that the amount and nature of the phytochemicals varies according to the season and geographical location [56].

In this regards, the investigation was carried out to check Antibiotic Sensitivity of *Staphylococcus aureus* which was isolated from raw milk samples of mastitic and healthy dairy cattle, by using extracts of wild plant species viz., *Prosopis juliflora, Cassia occidentalis and Tephrosia purpurea* plants. The present research work was carried out to investigate and find the properties of the extracts of *Prosopis juliflora, Cassia occidentalis and Tephrosia purpurea* under control laboratory conditions.

# 2. Material and Method

## 2.1. Collection and Extract Preparation of Plant Species

*Cassia occidentals, Prosopis juliflora and Tephrosia purpurea* species were collected from farms and grassy areas near Krishi Vigyan Kendra (KVK), Latur. Leaves of Disease Free plants were selected for test. Database of these plants are shown in (Table. 2.1)

50gm of each species was taken by weighing it. Collected leaves of *Prosopis juliflora, Cassia occidentalis and Tephrosia purpurea* were washed thoroughly and then oven dried till complete dehydration. The dried leaves then pulverized into a granular form by grinding in a mixer. Leaves were pulverized separately into powder form. 200ml of water added in 50gm powder of each species, were mixed properly by stirring for 24 hrs in stirrer. After stirring, all extracts were filtered by using Whatman no. 1 filter paper. The filtrate was concentrated in oven at 35°C for 20 mins. and stored at 4°C until further use.

**Table 2.1. Data of tropical medicinal plants**

| Scientific Names | Family | Habit | Part used | Properties | Reference |
|---|---|---|---|---|---|
| *Cassia occidentalis* | Caesalpiniaceae | Under shurb | Leaves, stem, seed | Diuretic, Antibacterial | [22] |
| *Prosopis juliflora* | Leguminosae | Shurb | | | [22] |
| *Tephrosia purpurea* | Papilionaceae | Pennial herb | Root, Leaves, Seed oil | Anthelmintic, Alexipharma | [22] |

## 2.2. Sample Collection of *S. Aureus*

Milk sample was collected from different dairies of Latur dist. between May-June 2019 and was collected in a 10ml sterile glass bottle. Sample was transported to Microbiology Research laboratory of Dayanand Science College, Latur, and stored at 4°C till the use.

## 2.3. Isolation, Identification and Characterization of *S. Aureus*

Isolation of *S. aureus* was attempted according to Singh and Prakash [12] with slight modification in methodology. Enrichment was carried out in Peptone Water (PW) enrichment broth. 10 ml or 10 g sample was homogenized with 90 ml sterile enrichment broth peptone water and enriched for 24 hrs at 37°C. A loopful of sample from enrichment were streaked on Blood agar plates and incubated at 37°C for about 24hrs (Figure 2.3a). Colony Characters and gram's nature of *Staphylococcus aureus* was identified by performing gram's staining. The appearance of jet black colonies surrounded by white halo was considered to be presumptive *S. aureus* (Figure 2.3b). After identification and confirmation, the pure culture was prepared by streaking it on Nutrient agar plates and incubated for 24 hours at 37°C. The plates were further characterized by biochemical tests.

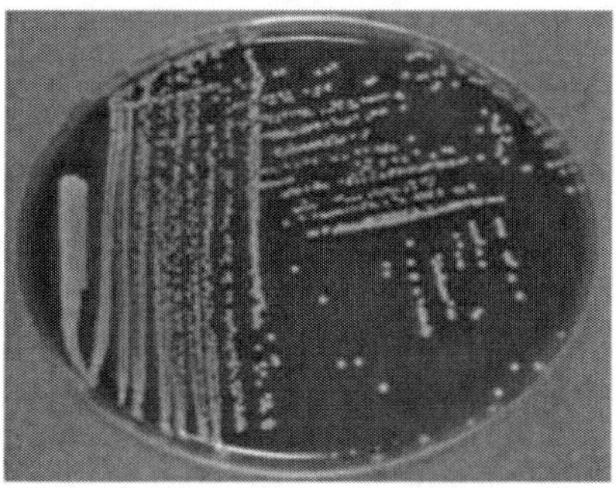

Figure 2.3a. *S. aureus* colonies on Blood Agar.

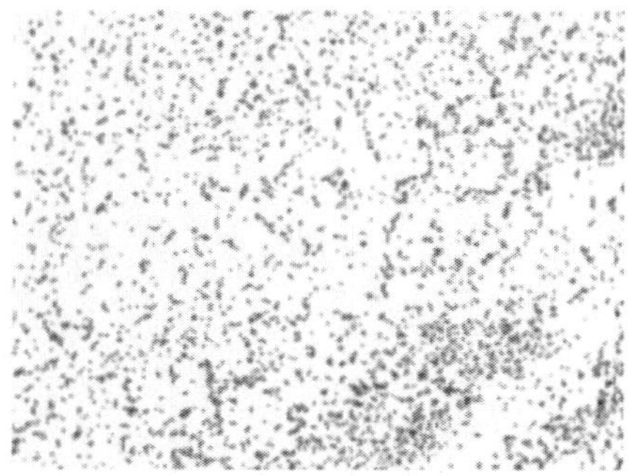

Figure 2.3b. Gram +ve cocci of *S. aureus*.

### *2.3.1. Morphological Studies*

The smear of isolate was prepared on clean grease free glass slide and heat fixed. The smear was stained with Grams staining method. The stained smear was observed under 100X or under oil immersion. The Gram positive, spherical cells arranged in irregular clusters resembling to bunch of grapes was observed under microscope (Table 2.3.1).

**Table 2.3.1. Morphological characters of *S. aureus***

| Sr. No | Test Performed | Results |
|---|---|---|
| 1 | Isolation of *Staphylococcus aureus* | white, circular convex, entire colonies On nutrient agar was seen |
| 2 | Pure culture | By quadrant method of streaking, Single isolated colonies were found on Nutrient agar |
| 3 | Colony morphology | Circular and pinhead colonies were observed on Trypticase soy agar, |
| 4 | Gram staining | Gram +ve 'cocci' |
| 5 | Motility By hanging drop method | bacteria was found to be motile. |
| 6 | Endospore staining | Under oil immersion, Non endospore formers were observed |

*2.3.2. Biochemical Viability*

TO Identify and confirm *Staphylococcus aureus*, various Biochemical tests were performed. The tests performed are Carbohydrate fermentation test, Coagulase test, Catalase test, Oxidase test and D-mannitol fermentation and results are tabulated in Table. 2.3.2.

**Table 2.3.2. Biochemical viability of isolate of *S. aureus***

| Sr. No | Biochemical Test Performed | Results | +/-ve |
|---|---|---|---|
| 1 | Catalase test | Effervesces were observed in the isolate. | +ve |
| 2 | Carbohydrate fermentation test | Isolate observed to be acid Positive | +ve |
| 3 | Coagulase test | Visible clumping was observed in the culture. | +ve |
| 4 | Oxidase test | No Oxidase Reduction observed in the Culture | -ve |
| 5 | D-mannitol fermentation Test | Mannitol Fermentation to be observed | +ve |

## 2.4. Growth Rate at 37°C and 45°C

The effect of varied temperature levels on growth of *Staphylococcus aureus* was studied and the data obtained is presented in (Table 2.4). Isolate of *Staphylococcus aureus* was tested at temperature 37°C and 45°C on NA agar medium. The data clearly indicated that the temperature of

37°C was found optimum for the growth of the pathogen as significantly the number of colonies were observed was maximum at this temperature. Isolate grew well at 37°C but no growth was observed at temperature of 45°C. Growth of *Staphylococcus aureus* started in 24hrs after incubation. Maximum growth was observed after 120 hrs of incubation at 37°C. Little growth was observed at ±48hrs of incubation at 45°C but, there were no growth after 48hrs upto 120hrs.

**Table 2.4. Growth rate of *Staphylococcus aureus* isolate**

| Sr. No. | Isolate | Growth rate | | | | | | | |
|---|---|---|---|---|---|---|---|---|---|
| | | 37°C | | | | 45°C | | | |
| | | 48hrs | 72hrs | 96hrs | 120hrs | 48hrs | 72hrs | 96hrs | 120hrs |
| 1 | *Staphylococcus aureus* | + | ++ | +++ | ++++ | + | - | - | - |

(Colony growth: '-': no growth, '+': less growth, '++': moderate growth, '+++': maximum growth)

## 2.5. Antibiogram Pattern of the Isolated *S. Aureus* to Plant Extracts

After conformation of *Staphylococcus aureus*, the antibacterial activity of plant extract was preliminarily screened by well diffusion assay. Nutrient agar medium was mixed with 20 µL bacterial strain ($1x10^8$ cfu/ml) and then spread plated on sterile petri plates. The wells of 6 mm diameter were made in the agar plates. Each plant extract was tested for antibacterial activity by adding 40 µL of extracts in different concentrations *viz*., 50 and 100 mg $ml^{-1}$. The experiment was repeated thrice. The plates with bacterial culture were then incubated at ±37°C for 24 hour. Subsequently, the plates were examined for zone of inhibition (ZOI) and diameter was measured in mm after subtracting well diameter.

To determine Minimum inhibitory concentration (MIC), required quantity of extracts were added in to the Petone water. Extracts were added in different 10 ml Peptone water tubes to set concentrations of 50 and 100 mg $ml^{-1}$. In each test tube 0.1 ml of standardized inoculum ($1x10^8$ cfu/ml) was added. Two control tubes were maintained for each test batch namely extract control (tube containing plant extract and Peptone Water without inoculum) and other as a control i.e., tube containing Peptone water and inoculum). The test tubes were incubated at 37°C for 24 hour. Turbidity was measured by using Terbidity meter and calculated in Nephelometric Turbidity Units (NTU). The lowest concentration (highest dilution) of plant extract that produced no visible growth (no turbidity) recorded as minimum inhibitory concentration.

# 3. Results

## 3.1. Well Diffusion Assay

The preliminary screening of selected three plant extracts against the *Staphylococcus aureus* was done using well diffusion method. The zone of inhibition greater than 5 mm diameter is found to be having significant activity against particular bacteria. The aqueous extracts of different concentrations were sensitive against *Staphylococcus aureus* and the data tabulated in Table 3.1.

**Table 3.1. Antibacterial sensitivity of plant extracts showing ZOI**

| Sr. No | Plant species | Conc. mg $ml^{-1}$ | Zone of inhibition (mm)* |
|---|---|---|---|
| 1 | *Cassia occidentalis* | 50 | 3.45 |
| | | 100 | 11.23 |
| 2 | *Tephrosia purpurea* | 50 | 9 |
| | | 100 | 14.01 |
| 3 | *Prosopis juliflora* | 50 | 6 |
| | | 100 | 12 |

The aqueous extract of *Tephrosia purpurea* was highly sensitive against *Staphylococcus aureus* tested at different concentrations. The same extract was much effective at low concentration of 50 mg $ml^{-1}$ i.e., (9 mm) ZOI. The extract of *Prosopis juliflora* have shown the zone of inhibition about (12 mm) and (6 mm) at 100 and 50 mg ml-1 respectively. The extracts of

*Cassia occidentalis* exhibited the lowest antibacterial activity of (3.45 mm) ZOI at 50 mg $ml^{-1}$ while (11.23 mm) ZOI at 100 mg $ml^{-1}$ .

**Table 3.2. Antibacterial activity of plant extracts showing MIC**

| Sr. No | Plant species | Conc. mg $ml^{-1}$ | Minimum Inhibitory Concentration(MIC) in NTU |
|---|---|---|---|
| 1 | *Cassia occidentalis* | 50 | 0.13 |
| | | 100 | 1.243 |
| 2 | *Tephrosia purpurea* | 50 | 0.98 |
| | | 100 | 2.001 |
| 3 | *Prosopis juliflora* | 50 | 0.75 |
| | | 100 | 1.891 |

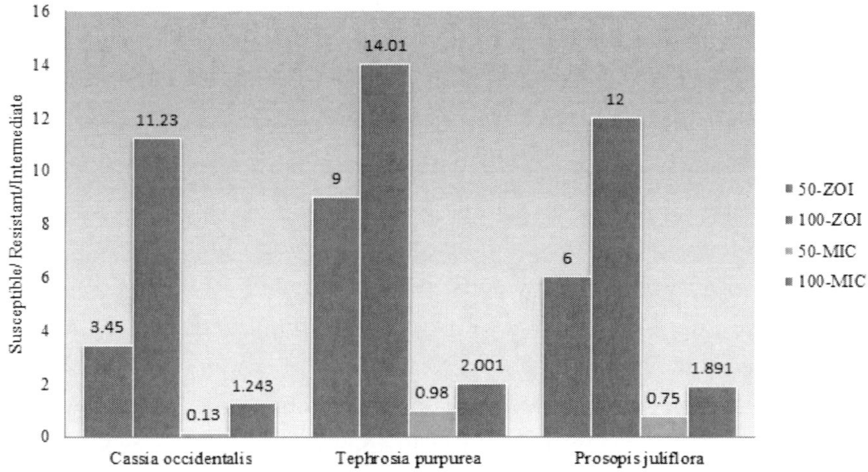


Figure 3. Graphical representation of ZOI and MIC.

## 3.2. Minimum Inhibitory Concentration

The antibacterial activity of *Tephrosia purpurea and Prosopis juliflora* extracts were found to be high against *S. aureus*. With analyzing with turbidity in peptone water, Minimum Inhibitory Concentration were determined. Strongest Inhibition was shown by *Tephrosia purpurea at* both the concentrations viz., 50 and 100mg $ml^{-1}$. Complete turbidity was observed in tube of *Cassia occidentalis* 50 mg ml-1 concentration while slight turbidity occur at 100 mg ml-1 (Table 3.2).

## 4. Discussion

*Staphylococcus aureus* is one of the common infective agent in hospitals and is most sensitive towards new born babies, old and malnourished persons, patients with diabetes, chronic diseases and surgical patients [57]. *Staphylococcus aureus* was found to be a frequent cause of burns and wound sepsis [58]. A study [59] at Ilorin, Nigeria reported wound infections of 38% as the highest frequency of *S. aureus* isolates. Some years ago, Cloxacillin was highly recommended in staphylococcal infection in view of excellent in vitro sensitivity results. This could be seen from the reports at Ilorin, Nigeria [59] with sensitivity of 78% and at Owerri, Nigeria [60] with 85.4% sensitivity. Neverthless, the results obtained are showing variance with the current trend in *S. aureus* susceptible to Cloxacillin as could be seen from the results of the present study and the reports of some other researchers [61, 62] who all observed very low sensitivity percentage to Cloxacillin. This became more common with the advent of Methicillin resistance as they belong to the same class of antibiotics.

The low sensitivity percentage of *S. aureus* observed against the drugs like, Chloramphenicol, Tetracycline, Cotrimoxazole and Penicillin was in agreement with the data published by some researchers [59, 62] from Nigeria and [63] in Eritrea. The high level of resistance could be associated with earlier sensation of these drugs to isolates which may produce much efficient resistance development. There is highest level of antibiotic abuse in this environment getting increased from self-medication which is most often related with inadequate dosage which leads in failure to comply for treatment [64], and availability of millions of antibiotics for people across the markets with or without any prescription [65].

In this study, it is found that the extracts of *Tephrosia purpurea*, *Prosopis juliflora* and *Cassia occidentalis* great antimicrobial activity against disease causing agent *S. aureus*. In the present study, *S. aureus* have shown an excellent susceptibility to *Prosopis juliflora*, *Tephrosia purpurea* and *Cassia occidentalis.* From the results, it has been indicated that the extracts of *Tephrosia purpurea*, *Prosopis juliflora* and *Cassia occidentalis* are having impressive potential to be used in the management and cure of *S. aureus* borne diseases. Further phytochemical analysis is required to identify the active components of plant extracts showing antimicrobial effect.

## CONCLUSION

Plant origin Bactericides of can be one the prime step to disease management because of their eco-friendly nature [66]. The synthetic drugs of plant origin are of greater advantage to user, the public and radical environmentalist. Laboratory study and analysis of plant extracts has given an encouraging results, indicating their potential use in the management of disease caused by *S. aureus* species. Plant extracts resulted in antibacterial activity is potential to use in the management of plant diseases as an alternative to chemical antibiotics. Further phytochemical analysis is required to identify the bioactive compounds responsible for antibacterial activity. Yet the results gained by these methods provide simple data that make it possible to differentiate extracts according to their antioxidant potential.

## ACKNOWLEDGMENT

The author wants to thanks Dr. Jaiprakash Dargad (Principal, Dayanand Science College, Latur) for providing Research Laboratory and Chemicals required during work. The author wants to extend her thanks to Dayanand Education Society for providing Research funding to author. Special Thanks to Dr. R. S. Shinde for his impressive collaboration with author. Also, the author heartily acknowledge to all supporting teachers and laboratory assistants of Dayanand Science College, Latur.

## References

[1] Loir, YL; Baron, F; Gautier, M. Staphylococcus aureus and food poisoning. *Gen. & Mol. Res.*, (2003), 2(1), 63-76.

[2] O'Riordan, K; Lee, JC. Staphylococcus aureus capsular polysaccharides. *Clin. Microbiol. Rev.*, (2004), 17, 218–234.

[3] Witte, W; Strommenger, B. Klare; Werner, IG. Antibiotic-resistant nosocomial pathogens. Part I: Diagnostic and typing methods. *Bundesgesundheitsblatt Gesundheitsforschung Gesundheitsschutz*, 47, (2004), 352–362.

[4] Shuval, H. Estimating the global burden of thalassogenic diseases: Human infectious diseases caused by wastewater pollution of the marine environment. *J. Water Health*, 2003, 1, 53–64.

[5] Yau, V; Wade, TJ; deWilde, CK; Colford, JM. Jr. Skin-related symptoms following exposure to recreational water: A systematic review and meta-analysis. *Water Qual. Expos. Health*, 2009, 1, 79–103.

[6] Moore, PCL; Lindsay, JA. Genetic variation among hospital isolates of methicillin-sensitive Staphylococcus aureus: Evidence for horizontal transfer of virulence genes. *J. Clin. Microbiol.*, 2001, 39, 2760–2767.

[7] Boyd, EF; Brüssow, H. Common themes among bacteriophage-encoded virulence factors and diversity among the bacteriophages involved. *Trends Microbiol.*, 2002, 10, 521–529.

[8] Robinton, ED; Mood, EW. A quantitative and qualitative appraisal of microbial pollution of water by swimmers: A preliminary report. *J. Hyg.*, 1966, 64, 489–499.

[9] Charoenca, N; Fujioka, R. Association of staphylococcal skin infections and swimming. *Water Sci. Technol*, 1995, 31, 11–17.

[10] WHO. Coastal and Fresh Waters. In Guidelines for Safe Recreational Water Environments, World Health Organization: Geneva, Switzerland, 2003, Volume 1.

[11] Papadakis, JA; Mavridou, A; Richardson, SC; Lampiri, M; Marcelou, U. Bather-related microbial and yeast populations in sand and seawater. *Water Res.*, 1997, 31, 799–804.

[12] Onile, BA; Odugbemi, TO; Nwofor, C. Antibiotic susceptibility of Bacterial agents of Septicemia in Ilorin. *Nig Med Pract.*, 1985, 9(4), 16-18.

[13] Wheat, LJ; Kholer, RB; White, A. Treatment of nasal carriers of coagulase positive staphylococci. In: Maibach, H. I, Aly, R.

[14] Tuazon, CV; Perez, A; Kishaba, T; Sheagren, JN. Staphylococcus aureus among Insulin Injecting diabetic patients: An increased carrier rate. *JAMA.*, 1975, 231, 1272.

[15] Yu, VL; Goetz, A; Wagener, M. Staphylococcus aureus Nasal carriage and infection inpatients on haemodialysis: Efficacy of antibiotic prophylaxis. *New Engl J Med.*, 1986, 315, 91-96.

[16] Luzar, MA; Coles, GA; Faller, B; Slingeneyer, A; Dah Dah, G; Brait, C; Wone, C; Knefati, Y; Kessler, M; Peluso, Y. Staphylococcus aureus nasal carriage and infection in patients on continuous ambulatory peritoneal dialysis. *N Engl J Med.*, 1990, Feb 22, 322(8), 505-9.

[17] Tuazon, CV; Sheagren JN. Increased rate of carriage of Staphylococcus aureus among narcotics addicts. *J Infect Dis.*, 1974, 129, 725.

[18] Kirmany, N; Tuazon, CV; Alling, D. Carriage of Staphylococcus aureus among patients receiving allergy injections. *Ann allergy.*, 1980, 45, 235-237.

[19] Raviglone, MC; Marvius, P; Pabloz-Mendel, A; Battan, R; Ottuso, P; Taranta, A. High Staphylococcus aureus nasal carriage rate in patients with acquired immunodeficiency syndrome (AIDS) related complex. *Am J Infect Control.*, 1990, 18, 64-69.

[20] Kloos, WE. Staphylococcus. In: *Topley and Wilson's Microbiology and Microbial Infections*, London: Edward Arnold, Vol. 2, 9th edn, 1998, 602-611.

[21] Montefiore, D; Rotimi, VO; Adeymi-Doro, FAB. The problem of bacterial resistance to antibiotics among strains isolated from hospital

patients in Lagos and Ibadan, Nigeria. *J Antimicrob Chemother.*, 1989 Apr, 23(4), 641-51.

[22] Joy, PP; Thomas, J; Samuel Mathew, Baby P. Skaria. *Medicinal Plants, Detabase on tropical plants.*, 1998, 186-210.

[23] Richards, C; Emori, TG; Edwards, J; Fridkin, S; Tolson, J; Gaynes, R. Characteristics of hospitals and infection control professionals participating in the National Nosocomial Infections Surveillance System 1999. *Am J Infect Control.*, 2001, 29, 400–403.

[24] Moran, GJ; Krishnadasan, A; Gorwitz, RJ; Fosheim, GE; McDougal, LK; Carey, RB; Talan, DA. Methicillin-resistant S. aureus infections among patients in the emergency department. *N Engl J Med.*, 2006, 355, 666–674.

[25] Jevons, MP. "Celbenin"-resistant staphylococci. *BMJ.*, 1961, 1, 124–125.

[26] Waness, A. Revisiting methicillin-resistant Staphylococcus aureus infections. *J Glob Infect Dis.*, 2010, 2, 49–56.

[27] Katayama, Y; Ito, T; Hiramatsu, K. A new class of genetic element, staphylococcus cassette chromosome mec, encodes methicillin resistance in Staphylococcus aureus. *Antimicrob Agents Chemother.*, 2000, 44, 1549–1555.

[28] Li, S; Skov, RL; Han, X; Larsen, AR; Larsen, J; Sørum, M; Wulf, M; Voss, A; Hiramatsu, K; Ito, T. Novel types of staphylococcal cassette chromosome mec elements identified in clonal complex 398 methicillin-resistant Staphylococcus aureus strains. *Antimicrob Agents Chemother.*, 2011, 55, 3046–3050.

[29] Shore, AC; Deasy, EC; Slickers, P; Brennan, G; O'Connell, B; Monecke, S; Ehricht, R; Coleman, DC. Detection of staphylococcal cassette chromosome mec type XI carrying highly divergent mecA, mecI, mecR1, blaZ, and ccr genes in human clinical isolates of clonal complex 130 methicillin-resistant Staphylococcus aureus. *Antimicrob Agents Chemother.*, 2011, 55, 3765–3773.

[30] Farnsworth, NR; Akerele, O; Bingel, AS; Soejarto, DD; Guo, Z. Medicinal plants in therapy. *Bull World Health Organ.*, 1985, 63(6), 965-81.

[31] Pasiecznik, NM; Harris, PJC; Smith, SJ. Identifying tropical *Prosopis* species: a field guide. Coventry: HDRA Publishing, 2004.
[32] González-Lamothe, R; Mitchell, G; Gattuso, M; Diarra, MS; Malouin, F; Bouarab, K. Plant antimicrobial agents and their effects on plant and human pathogens. *Int J Mol Sci.*, 2009, 10(8), 3400-19.
[33] Ji, HF; Li, XJ; Zhang, HY. Natural products and drug discovery. Can thousands of years of ancient medical knowledge lead us to new and powerful drug combinations in the fight against cancer and dementia? *EMBO Rep.*, 2009, 10(3), 194-200.
[34] Singh, S; Swapnil, Verma SK. Antibacterial properties of alkaloid rich fractions obtained from various parts of *Prosopis juliflora. Int J Pharma Sci Res.*, 2011, 2(3), 114-20.
[35] Burkart, A. A monograph of the genus *Prosopis* (Leguminosae subfam. Mimosoideae). *J Arnold Arbor.*, 1976, 57, 219–49.
[36] Mazzuca, M; Kraus, W; Balzaretti, V. Evaluation of the biological activities of crude extracts from Patagonian *Prosopis* seeds and some of their active principles. *J Herb Pharmacother.*, 2003, 3(2), 31-37.
[37] Tene, V; Malagón, O; Finzi, PV; Vidari, G; Armijos, C; Zaragoza, T. An ethnobotanical survey of medicinal plants used in Loja and Zamora-Chinchipe, Ecuador. *J Ethnopharmacol.*, 2007, 111, 63–81.
[38] Murthy, MSR; Srinivasan, M. *Indian J Pharmacol.*, 1993, 25, 34-36, 1-5.
[39] Santram Lodhi, Rajesh Singhpawer, Alok Pal Jain, Singhai, AK. *J Ethnopharmacol* ., 2006, 108, 204-210.
[40] Kumar, A; Dutta, M; Bhatt, TK; Dalal, DS. *Indian Vet J.*, 1997, 74, 424.
[41] Sankaran, JR. *The Antiseptic.*, 1980, 77, 643.
[42] Khatri, A; Garg, A; Agrawal, S. *J Ethnopharmacol.*, 2009, 122, 1-5.
[43] Kumar, GS; Jayaveera, KN; Kumar, CK; Sanjay, UP; Vrushabendra, BM; kumar, DV. *Trop J Pharma Res.*, 2007, 6, 717-723.
[44] Saxena, BN; Dubey, DN; Nair, AL. *Defence Sci J.*, 1974, 24, 43-48.
[45] Swathi, D; Sujatha, D; Bharathi, K; Prasad, KVS. *phramacognosy magazine.*, 2008, 4, 206-211.

[46] Pavana, P; Sethupathy, S; Manoharan, S. *Indian J Clinical Biochem.*, 2007, 22, 77-83.

[47] Sonawane, IL; Nirmal, S; Dhasade, V; Rub, R; Mandal, S. *Int J Pharma Sci Res.*, 2010, 1, 57-60.

[48] Gupta, AK. Quality standards of Indian medicinal plants. I sted. *New Delhi: Indian council of medical research*, 2003, 193.

[49] Jain, SC; Sharma, RA; Jain, R; Mittal, C. Antimicrobial screening of *Cassia occidentalis* L *in vivo* and *in vitro*. *Phytotherapy Research*, 1998, 12, 200-204.

[50] Saganuwan, AS; Gulumbe, ML. Evaluation of *in vitro* antimicrobial activities and phytochemical constituents of *Cassia occidentalis*. *Animal Research International*, 2006, 3, 566-569.

[51] Tona, L; Ngimbi, NP; Tsakala, M; Mesia, K; Cimanga, K; Apers, S; De Bruyne, T; Pieters, L; Totté, J; Vlietinck, AJ. Antimalarial activity of 20 crude extracts from nine African medicinal plants used in Kinshasa Congo. *Journal of Ethnopharmacology*, 1999, 68, 193-203.

[52] Jafri, MA; Subhani, MJ; Javed, K; Singh, S. Hepatoprotective activity of leaves of *Cassia occidentalis* against paracetamol and ethyl alcohol intoxification in rats. *Journal of Ethnopharmacology*, 1999, 66, 355-61.

[53] Sharma, N; Trikha, P; Athar, M; Raisuddin, S. *In vitro* inhibition of carcinogen induced mutagenicity by *Cassia occidentalis* and *Emblica officinalis*. Drug and Chemical Toxicology, 2000, 23, 477-84.

[54] Tona, L; Cimanga, RK; Mesia, K; Musuamba, CT; De Bruyne, T; Apers, S; Hernans, N; Miert, SV; Pieters, L; Totté, J; Vlietinck, AJ. *In vitro* antiplasmodial activity of extracts and fractions from seven medicinal plants used in the Democratic Republic of Congo. *Journal of Ethnopharmacology*, 2004, 93, 27-32.

[55] Sharma, N; Trikha, P; Athar, M; Raisuddin, S. *In vitro* inhibition of carcinogen-induced mutagenicity by *Cassia occidentalis* and *Emblica officinalis. Drug and Chemical Toxicology*, 23, 2000, 477-84.

[56] Yadav, JP; Arya, V; Yadav, S; Panghal, M; Kumar, S; Dhankhar, S, *Cassia occidentalis*: A review on its ethnobotany, phytochemical and

pharmacological process. *Fitoterapia*, 2009. doi:10.1016/j.fitote.2009.09.008.

[57] Tuo, P; Montobbio, G; Vallarino, R; Tumolo, M; Calero, MG; Massone, MA. Nosocomial Staphylococci in a neonatal and pediatric intensive care unit. *Pediatr Med Chir.*, 1995 Mar-Apr, 17(2), 117-22.

[58] Emmerson, M. Nosocomial Staphylococcal outbreak. *Scand J Infect Dis Suppl.*, 1994, 93, 47-54.

[59] Ndip, RN; Ebah, LME; Onile, BA. Antibiogram of Staphylococcus aureus from clinical Syndromes in Ilorin, Nigeria. *J Med Lab Sci.*, 1997, 6, 24-26.

[60] Uwazuoke, JC; Aririatu, LE. A survey of Antibiotic resistant Staphylococcus aureus strains from clinical sources in Owerri. *J Appl Sci Environ Managt.*, 2004, 8(1), 67-68.

[61] Amadi, ES; Ikeagwu, IJ; Iroha, IR. Antibiotic sensitivity pattern of Staphyloccoccus aureus in Abakiliki, Nigeria. *Pak J Med Sci.*, 2008, 24(2), 231-235.

[62] Obiazi, HAK; Nmorsi, OPG; Ekundayo, AO; Ukwuandu, NCD. Prevalence and antibiotic susceptibility pattern of Staphylococcus aureus from clinical isolates grown at 37 and 44°C from Irrua, Nigeria. *Afr J Microbiol Res.*, 2007, 57-60.

[63] Naik, D; Teclu, A. A study on antimicrobial susceptibility pattern in clinical isolates of Staphylococcus aureus in Eritrea. *Pan African Medical Journal.*, 2009, 3, 1.

[64] Odugbemi T. The use and abuse of antibiotics. *Nig Med Pract.*, 1981, 1(1), 4-8.

[65] Paul, MO; Aderibie, DA; Sule, CZ. Antimicrobial sensitivity patterns of hospital and non hospital strains of Staphylococcus aureus isolated from nasal carriers. *J Hyg Camb.*, 1982, 89, 253-260.

[66] Bolkan, HA; Reinert, WR. *Development and implementing IPM strategies to assist farmers, an industry approach, plant Disease.* 1994, 78, 545-550.

[67] Saga, T; Yamaguchi, K. History of antimicrobial agents and resistant bacteria. *Jpn Med Assoc J.*, 2009, 52(2), 103–08.

In: Innovations and Challenges in Modern … ISBN: 978-1-53618-181-4
Editor: A. K. Haghi 

*Chapter 5*

# EFFECT OF LIQUID FORMULATION OF *AZOTOBACTER VINELANDII* ON GERMINATION OF *VIGNA RADIATE* (MUNG BEANS)

***Aishwarya A. Andhare and Ravindra S. Shinde****
Department of Microbiology, Biotechnology and Chemistry,
Dayanand Science College, Latur, Maharashtra, India

## ABSTRACT

The purpose of the present research were to study the population dynamics of *Azotobacter vinelandii* at periodic interval and its effect on growth indices and nitrogen intake by *Vigna radiate* (Mung Beans) crop. In all there were three treatments including seed treatment with liquid *A.vinelandii 1*, *A. vinelandii 2*, *A. vinelandii 3* in concentration of 50 ml/kg of each. Result revealed that the application of liquid formulation of *Azotobacter vinelandii* shown significantly increased crop growth and improved soil biochemical properties over control. The higher per cent of seed germination (87.26) recorded in seed treated with liquid *A. vinelandii*

* Corresponding Author's Email: rshinde.33381@gmail.com.

*3* while the lower seed germination per cent (83.97) was recorded in seed treated with liquid *A.vinelandii 1*. The seed treated with liquid *Azotobacter vinelandii* recorded significant increase in respect of Length of Leaves, Speed of Emergence (SE) and seedling vigor Index (SVI). More SE and SVI was observed by *A.vinelandii 3* (SE= 86.79% and SVI= 83.53) while lowest SE and SVI was recorded by *A.vinelandii 1* (SE= 81.43% and SVI= 80.82). Considering all these objectives, it could be estimated that *Azotobacter vinelandii* have improved soil biochemical properties which may ultimately influence the growth of *Vigna radiate* (Mung Beans) crop.

**Keywords:** liquid formulation, phyto-pathogens, seed germination, *Vigna radiate,* rhizospheric microbes

## 1. Introduction

Mung bean [*Vigna radiata* (L.)] is a tropical legume plant. Till the year of 2009, annual production of *Vigna radiata* crop was about 35,000 tons generating over 1,779.93 million Rs in exports to India, Japan, Philippines and to the USA. Based on the increased current demand of this crop, there is demand to annually increase the size of the mung bean industry to 100,000 tons. From the history, poor famer describes mung bean as a low yielding and high risk crop after a winter cereal [1] has also restricted the increase of the crop into new lands and planting areas [2].

The recently released mung bean cultivar Crystal has about 20% yield advantage when compared to the first germplasm introduction of varieties such as Emerald and White Gold in Australia. The cultivation of Crystal therefore is resulting in higher profitability boosting growers' confidence and improvement in the perception about the crop. However, its yield potential is still less than that of other summer legumes. Yield of up to 5 t/ha of newly released peanut and soybean cultivars of comparable maturity are being realized in good rainfall years in Australia. This suggests that realisation of high commercial yield from legumes is possible. In soybean and peanut, selection for a high partitioning trait has apparently contributed to the development of higher yield potential [2-6]. The possibility of improving partitioning traits has, however, not received similar attention in mungbean.

'Biofertilizer' is a substance which contains living microorganism which when applied to seeds, plant surfaces or soil, colonizes the rhizosphere or the interior of the plant and promotes growth by increasing the supply or availability of primary nutrients to the host plant. Chemical fertilizers directly increase soil fertility by adding nutrients. However biofertilizer add nutrients through the natural processes of fixing atmospheric nitrogen, solubilizing phosphorus, and stimulating plant growth through the synthesis of growth promoting substances. Liquid biofertilizer production is an alternative solution to carrier based biofertilizers. It comprises aids to preserving organism, to delivering them to their targets and improves their activities. Liquid biofertilizers have a longer shelf life in comparison with lignite based biofertilizers [7]. By applying an appropriate liquid biofertilizer, the overall cost of production will be much lower as compared to traditional chemical fertilizers [8].

*Azotobacter* is commonly found in soil especially in rhizosphere and phyllosphere of plants. It is very effective for the improvement of crop productivity by improving soil fertility. It can fix nitrogen directly from the atmosphere that helps plants for better grain production. Besides, nitrogen fixation, *Azotobacter* also produces growth hormones *viz;* riboflavin, thiamine, nicotine, indole acetic acid and gibberellins [9-11]. *Azotobacter vinelandii* is a common Gram-negative soil bacterium that can fix atmospheric nitrogen under aerobic conditions. This characteristic differentiates it from many other nitrogen fixing bacteria that require anaerobic or microaerobic conditions to protect the oxygen-sensitive nitrogenase [12]. This feature also makes it an ideal candidate strain for potential co-culture with oxygen-producing phototrophs such as microalgae [13]. Additionally, *A. vinelandii* is considered as an ideal strain in potential biotechnology applications for the production of higher value bioproducts such as polyhydroxyalkanoates, which could serve as potential bioplastics [12]. *A. vinelandii* has also been reported to excrete a range of additional nitrogen compounds to serve various functions, including proteins involved in the production of alginate [14].

By analyzing the previous studies of different scientists of globe, the present work was carried out to study the Effect of liquid formulation of *A.*

*vinelandii* 1, *A. vinelandii* 2, *A. vinelandii 3* inoculation on growth and yield of *Vigna radiate* (Mung Beans).

## 2. Role of Azotobacter Spp. in Soil

### 2.1. Soil with *Azotobacter*

The appearance of *Azotobacter* in soils has beneficial effects on plants, but the abundance of these bacteria is related to many factors, soil physico–chemical (e.g., organic matter, pH, temperature, soil moisture) and microbiological properties [15]. Its abundance varies as per the depth of the soil profile [16-20]. *Azotobacter*ia are much more abundant in the rhizosphere of plants than in the surrounding soil and that this abundance depends on the crop species [21, 22].

### 2.2. Biological Nitrogen Fixation

Nitrogen is one of the important elements of major nutrients and Biomolecules such as nucleic acids, protein and chlorophyll. So, nitrogen supply to the plant will influence the amount of protein, amino acids, protoplasm and chlorophyll formed. Therefore, adequate supply of nitrogen is necessary to achieve high yield potential in crop.

The earth atmosphere contains approximately 78% of nitrogen which is either inert or in unavailable form. For every hectare of ground there are approximate 80000 tones of this unavailable nitrogen. In order to be converted to available form it needs to be fixed through either the industrial process or through Biological Nitrogen Fixation (BNF). Without these nitrogen fixers, the atmospheric nitrogen couldn't get involved in plants and life may not be existing.

Nitrogen (N) deficiency is frequently a major limiting factor for crops production [24]. Nitrogen is an essential plant nutrient, widely applied as Nitrogen fertilizer to improve yield of agriculturally important crops. An interesting alternative to avoid or reduce the use of Nitrogen fertilizers could be the exploitation of Plant Growth Promoting Bacteria (PGPB) capable of enhancing growth and yield of many plant species, several of agronomic and ecological significance.

*Azotobacter is* non symbiotic heterotrophic bacteria capable of fixing an average 20kg of Nitrogen per Hectare per year [15]. Bacterization helps to improve plant growth and to increase soil nitrogen through nitrogen fixation by utilizing carbon for its metabolism [24].

## 2.3. Seed Inoculation *Azotobacter* and Nutrient Uptake

Seed Inoculated with *Azotobacter* helps in uptake of N, P along with micronutrients like Fe and Zn, in wheat, these strains can potentially be used to improve wheat nutrition [25]. Seed inoculation of *Azotobacter* profoundly contribute to increase yield by supplying nitrogen to the crops. Inoculation of seeds with *Azotobacter chroococcum*increased carbohydrate and protein content of two corn varieties (Inra210 and Inra260) in greenhouse experiment [26]. There is increment in Maize biomass with the application of manure and *Azotobacter* [27]. In nitrogen–deficient sand, seed inoculation increased plant length, dry weight, and nitrogen content in addition to a significant increase in soil nitrogen [28]. It was found that *A. chroococcum*at concentration of 108cfu ml–1 increased seed germination of Cucumber. Seeds of wheat (*Triticum aestivum*) were inoculated with 11 bacterial strains of *A.chroococcum*, Research result showed that all *A. chroococcum*s trains had positive effect on the yield and N concentrations of wheat [29].

## 2.4. Role of Azotobacter in Growth Substances Production and Promotion

Besides, nitrogen fixation, *Azotobacter* produces, Thiomin, Riboflavin, Nicotin, Indol Acetic Acid and Giberalin. When *Azotobacter* is applied to seeds, seed germination is improved to a considerable extent. Brakel et al.,

[30], showed that *Azotobacter* produced Indol–3–Acetic Acid (IAA) when tryptophan was added to the medium. Hennequin et al., [31] found only small amounts of IAA in old cultures of *Azotobacter* to which no tryptophan was added.

Bacteria of the genus *Azotobacter* synthesize auxins, cytokinins, and GA–like substances, and these growth materials are the primary substance controlling the enhanced growth of tomato [32]. These hormonal substances, which originate from the rhizosphere or root surface, affect the growth of the closely associated higher plants. Eklund et al., [33] demonstrated that the presence of *Azotobacter chroococcum* in the rhizosphere of tomato and cucumber is correlated with increased germination and growth of seedlings. Puertas et al., [34] report that dry weight of tomato plants inoculated with *Azotobacter chroococcum* and grown in phosphate–deficient soil was significantly greater than that of non inoculated plants. Phytohormones (auxin, cytokinin, gibberellin) can stimulate root development. High Gibberillic acid production was detected in *Azotobacter* (71.42%) isolates. Higher phosphate solubilization was detected in the isolates of *Azotobacter* (74.28%) followed by *Pseudomonas* (63.00%) [35]. Gibberellins applied in small quantities to the soil or rosettes on the leaves and shafts of certain plant produces an increase in height. In grains such as wheat and corn, they also cause an increase in length of the leaves. In some cases they also increase the fresh weight and dry weight. However it doesn't produce any effect on the growth of the roots [36].

Brown et al., [37] have found that after treating tomato seeds or seedling roots with small amounts (0.5–0–01pg.) of commercially produced gibberellins GA 3, the plants responded in the same way as after treatments with 14–day cultures of *Azotobacter chroococcum*strain A 6. *Azotobacter chroococcum*strains isolated from the sugar beet rhizosphere were also shown to produce gibberelins; the growth of pea hypocotyl was equivalent to a GA3 concentration of 0.003–0.1μg/cm3 culture [38]. Similarly, Cytokinins are related to nucleic acids with precursor substances that act to stimulate cell division in vegetative growth areas.

These responses suggest that *Azotobacter* probably influences the development of plants by producing growth–regulating substances. Therefore, *Azotobacter* spp. is often regarded as a member of "Plant Growth Promoting Rhizobacteria (PGPR)." It was also concluded that *Azotobacter* inoculants have a significant promoting effect on growth parameters like root, shoot length and dry mass of bamboo and maize seedlings in vitroand in pot experiments.2 The dual inoculation of *A. chroococcum*and *P. indica* had beneficiary response on shoot length, root length, fresh shoot and root weight, dry shoot and root weight, and panicle number that affect growth of rice plant [39].

*Adathoda vasica*plants inoculated with *A. chroococcum* revealed significantly increased root nitrogen content compared to the control plants [40]. The *Azotobacters* which were inoculated with *Rhizophora* seedlings, increased significantly the average root biomass up to by 98.2%, the root length by 48.45%, the leaf area by 277.86%, the shoot biomass by 29.49% as compared to controls and they also increased the levels of total chlorophylls and carotenoids up to by 151.0% and 158.73%, respectively [41]. The *Azotobacter* inoculated crop show better growth rate. During rosette stage of canola the Crop Growth Rate (CGR) found highest i.e., 10–12% increment [42].

## 3. Materials and Methods

### 3.1. Collection of Soil Sample

Collection of the soil samples were done from different habitat of Latur District, Maharashtra, India on 19 July 2018. Selected habitats were Murud, Renapoor and Sawe wadi. 15 cm depth of soil was considered to be ideal place to collect soil and then soil was collected in sterile polyethylene bags. Samples were then transported to the Research Laboratory of Dayanand Science College, Latur and stored at 4°C until use.

### 3.2. Isolation and Identification of *A. vinelandii* from Different Soil Samples for *In-Vitro* Studies

Isolation of *A. vinelandii* 1, *A. vinelandii* 2 and *A. vinelandii* were done by a serial dilution method. $10^3$ dilution of each sample was prepared and labeled separately. One milliliter suspension from each tube was pipetted onto Ashbys agar media plates, was spread with spreader and incubated at 28°C for 3 days. The culture plates were observed daily and the characters were noted till the appropriate results to be taken. Each appeared colony that was considered to be one colony forming unit (cfu). After enumeration of Colony Forming Units (cfu), the colonies with different colony characters were isolated from the same plates and each uncommon colony was reisolated on another Ashbys agar media plates. Distinct Micro-morphological characteristics were observed for identification and confirmation, then the plates were stored at 4°C to avoid contamination. Two techniques, visual observation on old cultured petri plates and micro-morphological studies by observing slide culture, were adopted for identification of *A. vinelandii* 1, *A. vinelandii* 2, *A. vinelandii.* For visual observation, the isolates were grown on Ashbys agar media for 3-4 days.

### 3.3. Characterization and Identification of *A. vinelandii* Species Isolated from Different Habitats

Isolated *A.vinelandii* cultures were identified on the basis of Cultural characters, Morphological and Biochemical studies such as cell morphology and grams staining. For cultural characters the parameters were considered were viz., Color, shape, pigment formation on Ashbys agar medium while for biochemical characteristics differentiating fermentation sugars such as Sucrose, Lactose, Glucose, Mannitol were used to Identify and confirm *A.vinelandii* species.

### *3.3.1. Gram's Nature and Colony Characters of Different Isolates of A. Vinelandii*

All the selected three isolates are gram negative. Even if the isolates were of same Genus and species, the colony morphology, color, margins, elevation, consistency and density of those three different isolates were found slight different than each other (Table 3.3.1).

**Table 3.3.1. Gram's nature and colony characters on Ashbys agar medium**

| Isolates | Gram staining | Form | Color | Elevation | Margin | Consistency | Density |
|---|---|---|---|---|---|---|---|
| *A.Vinelandii 1* | -ve | Circular | Off white | Convex | Entire | Mucoid | Not transparent |
| *A.Vinelandii 2* | -ve | Circular | Creamy white | Flat | Entire | Highly Mucoid | Not transparent |
| *A.Vinelandii 3* | -ve | Circular | Creamy white | Convex | Entire | Highly Mucoid | Not transparent |

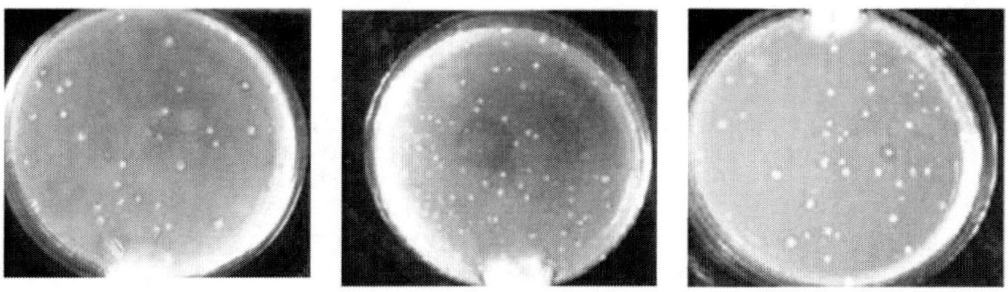

(1a) (1b) (1c)

Note: 1a= A.Vinelandii 1, 1b= A.Vinelandii 2, 1c= A.Vinelandii 3.

Figure 3.3.1a. Isolates of A.vinelandii on Ashbys agar medium.

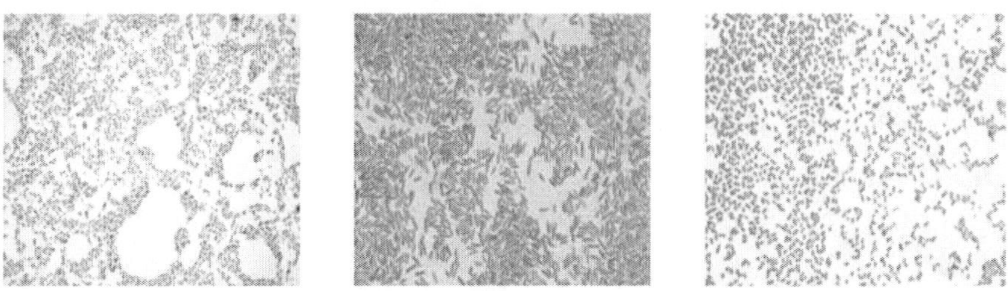

(2a) (2b) (2c)

Note: 2a= A.vinelandii 1, 2b= A.vinelandii 2, 2c= A.vinelandii 3.

Figure 3.3.1b. Microscopic study different isolates of *A.vinelandii.*

*3.3.2. Biochemical Analysis*

The isolates were studied for identification by using standard biochemical methods for *A.vinelandii* as directed in the Bergey's manual of systematic bacteriology (2001).

*3.3.2.1. Acid production from Sucrose, Lactose, Glucose, Mannitol, Starch*

The acid production by the isolates of *A.vinelandii* in test tubes were tested by using 10 ml of sterile glucose broth with bromo cresol purple as standard pH indicator. The tubes were incubated for 7 days at 30°C. Change in colour of medium to yellow was taken as positive for acid production. The media was sterilized for 15 min at 15 lb pressure in an autoclave. To these tubes, filter sterilized carbohydrates *viz.*, Sucrose, Maltose and Dextrose were added at 0.1/10ml concentration. The tubes were inoculated with 0.1 ml of 24 hrs old bacterial culture and incubated at normal room temperature for three days. Acid production was confirmed by change in the color of the medium (Table 3.3.2.1).

**Table 3.3.2.1. Acid production from Sucrose, Lactose, Glucose, Mannitol, Starch**

| **Isolates** | **Carbohydrates** | | | | |
|---|---|---|---|---|---|
| | **SUC** | **LAC** | **GLU** | **MAN** | **STR** |
| *A.Vinelandii 1* | A | A | A | A | -- |
| *A.Vinelandii 2* | A | A | A | A | -- |
| *A.Vinelandii 3* | A | A | A | -- | -- |

Note: SUC= Sucrose, LAC= Lactose= Lactose, GLU= Glucose, MAN= Mannitol, STR= Starch.

## 3.4. Screening of *A.vinelandii* Isolates for Temperature Tolerance

After Identification of *A.vinelandii* species, screening was done analyzing temperature tolerance. Screening for obtaining temperature tolerant *A.vinelandii* isolates shows interesting results, tabulated in Table 3.4.

**Table 3.4. Screening of *A.vinelandii* isolates for temperature tolerance**

| Sr. No | Isolates | Temperature (0C) | | | | | | |
|---|---|---|---|---|---|---|---|---|
| | | $28^0$C | $37^0$C | $40^0$C | $45^0$C | $50^0$C | $55^0$C | $60^0$C |
| 1 | *A.Vinelandii 1* | + | + | + | -- | -- | -- | -- |
| 2 | *A.Vinelandii 2* | + | + | -- | -- | -- | -- | -- |
| 3 | *A.Vinelandi 3i* | + | + | + | + | -- | -- | -- |

## 3.5. Sterilization of Soil

Around 20kg of soil was sterilized using Oven. Sterilized soil was poured in seven pots. Pots are named as A1, A2; B1, B2; C1, C2 and one as control (C0).

## 3.6. Preparation of Liquid Formulation

A Liquid Formulations of the test isolates of *A.vinelandii* were prepared seperatly from a 24hr old culture of the isolates on plates of Ashbys agar medium. The plates (9cm diameter) of *A.vinelandii* were flooded with 10ml of sterilized distilled water and properly shaken for a few minutes till the complete mixing. Then the suspensions were filtered through muslin cloth. After filtering the suspensions through double layer of cheese towel, the concentration of cells of *A.vinelandii* was determined by using a double ruled Nabuer's haemocytometer. The cell concentration of the filtrates were adjusted to $10^4$ *A.vinelandii* cells per ml using sterilized distilled water.Three *A.vinelandii* liquid formulations were prepared for seed treatement viz., LF1= *A.vinelandii 1,* LF2= *A.vinelandii 2,* LF3= *A.vinelandii 3.*

## 3.7. Seed Treatement

The seeds of *Vigna radiata* were collected from the Farm of black soil of Renapur, Dist. Latur of Maharashtra, India. The treatments listed below were carried out to determine the potential of Isolates of *A.vinelandii* as a seed treatment.

- **T1** 30 seeds were soaked in sterilized distilled water for 12 h.
- **T2** 10 seeds were treated with each Liquid Formulation (LF1, LF2, LF3) of the test *A.vinelandii* isolates at concentration of 50ml/kg for 24 h and named as S1, S2, S3.
- **T3** seeds were treated with 0.1% HgCl (Murcuric Chloride) solution as disinfectant for 30secs and washed in running water till complete removal of HgCl.

The treated seeds were air-dried for 1 day under existing environmental conditions before sowing as described. The experiments were conducted for 6 weeks during Months of July-August 2018 under favorable conditions. Six pots were prepared named as A1, A2; B1, B2; C1, C2. S1, S2 and S3 Seeds were subjected to a treatment (T1, T2, T3) were sowed in these six pots in such way that S1/S2/S3= 5 Seeds in A1/B1/C1 and 5 Seeds in A2/B2/C2. In control pot (C0), 5 seeds without any seed treatment were sowed. The field was irrigated at definite intervals of 3 days after sowing. After sowing, one plant from each plot was randomly uprooted on each week difference to determine germination percentage of seeds, Average Length of Leaves, seedling vigor Index (SVI) and Speed of Emergence (SE).

## 3.8. The Equations Used to Calculate Different Germination Indices

| Germination parameters | Equations | References |
|---|---|---|
| Germination percent (GP) | $\frac{\text{Number of germinated seeds at final count}}{\text{Total number of seeds sets for bioassay}} \text{ X } 100$ | Global method |
| Speed of emergence (SE) | $\frac{\text{Number of germinated seeds at the starting day of germiantion}}{\text{Number of germinated seeds at the final days of measurement}} \text{ X } 100$ | Modified from Islam et al. (2009) |
| Seedling vigour index (SVI) | $\frac{\text{Seedling length (mm) x Germination percent}}{100}$ | Islam et al. (2009) |

# 4. Result

During Seed treatment T2, Seeds were treated with each Liquid Formulation of *A.vinelandii* isolates viz., *A.vinelandii* 1, *A.vinelandii* 2, *A.vinelandii 3* for 24 h and sowed in six different pots. One pot was set for control study. The Test was conducted for Six weeks. Result revealed after each week. Result observed was increased germination indices.

## 4.1. Germination Percent (GP)

Germination percents were calculated after each week and compared with control. Results revealed increase in Germination percentages. The higher seed germination per cent recorded in seed treated with liquid *A.vinelandii 3* while the lower seed germination per cent was recorded in seed treated with liquid *A.vinelandii 1* (Table 4.1) and (Figure 4.1).

**Table 4.1. GP% during test weeks**

| Pot No. | GP% during Test Weeks | | | | | | Final GP% |
|---|---|---|---|---|---|---|---|
| | 1 | 2 | 3 | 4 | 5 | 6 | |
| A1, A2 | 24.21 | 41.76 | 59.52 | 71.21 | 79.52 | 83.97 | 83.97 |
| B1, B2 | 25.63 | 43.25 | 60.69 | 71.55 | 81.23 | 85.22 | 85.22 |
| C1, C2 | 27.5 | 44.16 | 61.26 | 72.82 | 85.27 | 87.26 | 87.26 |
| C0 | 21.23 | 41.52 | 57.3 | 68.11 | 71.25 | 78.19 | 78.19 |

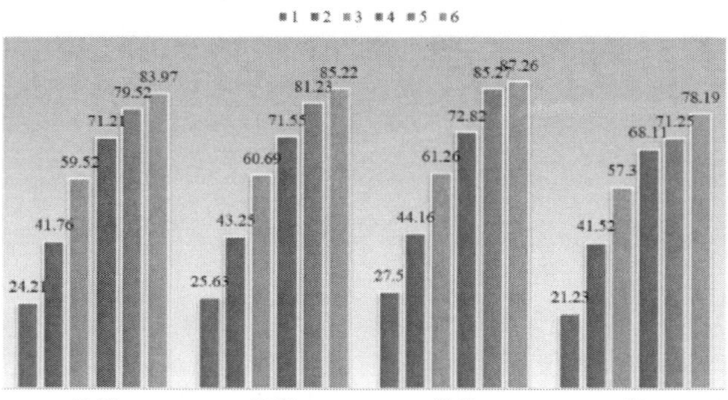


Note: A1, A2= S1 Seeds treated with *A.vinelandii 1*, B1, B2= S2 seeds treated with *A.vinelandii 2* and C1, C2= S3 seeds treated with *A.vinelandii 3,* C0= Control.

Figure 4.1. GP% during test weeks.

## 4.2. Seedling Vigor Index and Speed of Emergence

Speed of Emergence (SE) and Seedling Vigor Index (SVI) were calculated after 6 weeks. More SE and SVI was observed by *A.vinelandii 3* while lowest Seedling Vigor Index and Speed of Emergence was recorded by *A.vinelandii 1* (Table 4.2) and (Figure 4.2)

.

**Table 4.2. Germination indices during test weeks**

| Pot No. | Germination Indices during Test Weeks | |
|---|---|---|
| | Speed of Emergence (SE) % | Seedling Vigor Index (SVI) |
| A1, A2 | 81.43 | 80.82 |
| B1, B2 | 85.51 | 81.6 |
| C1, C2 | 86.79 | 83.53 |
| C0 | 79.29 | 75.67 |

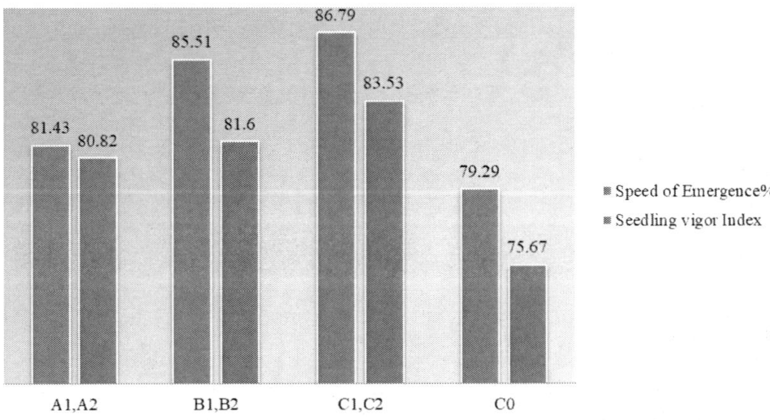


Note: A1, A2= S1 Seeds treated with *A.vinelandii 1*, B1, B2= S2 seeds treated with *A.vinelandii 2* and C1, C2= S3 seeds treated with *A.vinelandii 3,* C0= Control.

Figure 4.2. Germination Indices during test weeks.

## 4.3. Effect on Length of Leaves

The length of leaves was recorded at periodic interval after each 2 week of sowing with the help of metric scale. It was observed that the inoculation with liquid *A.vinelandii* significantly increased the length of leaves at different growth stages of *Vigna radiate* over the uninoculated control (Table 4.3) and (Figure 4.3).

**Table 4.3. Average length of leavesduring test weeks in cm**

| **Pot No.** | **Average length of leavesduring test weeks in cm** | | |
|---|---|---|---|
| | **2nd** | **4th** | **6th** |
| A1, A2 | 11.3 | 13.92 | 15.67 |
| B1, B2 | 12.32 | 14.12 | 15.93 |
| C1, C2 | 12.97 | 14.89 | 16.68 |
| C0 | 10.37 | 11.52 | 13.39 |

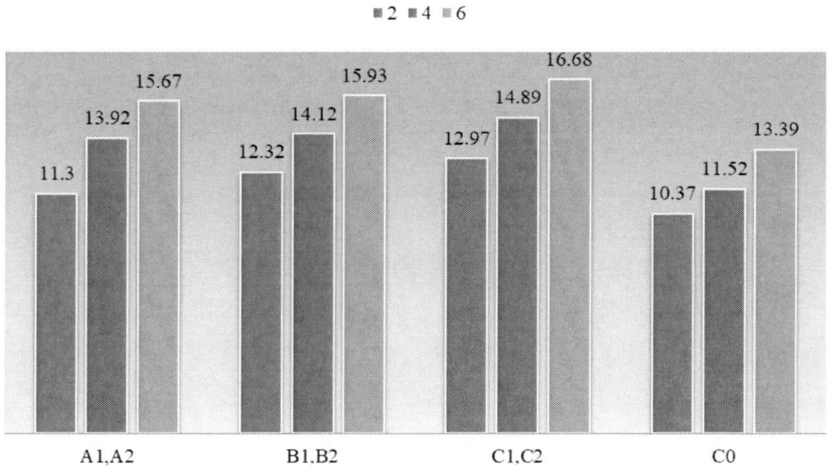


Note: A1, A2= S1 Seeds treated with *A.vinelandii 1*, B1, B2= S2 seeds treated with *A.vinelandii 2* and C1, C2= S3 seeds treated with *A.vinelandii 3,* C0= Control.

Figure 4.3. Average length of leaves during test weeks in cm.

## DISCUSSION

The present experiment was carried out with an objective to study and find the potential of liquid formulation of *A.vinelandii* on growth and yield of *Vigna radiata.* The results obtained by this study work are described herewith and analyzed with earlier research works. The results on seed germination shown statistically non significant over uninoculated control. However, marginal increase in seed germination in seed treatment with liquid *A.vinelandii*. When the seeds were treated with liquid *A.vinelandii 3,* seed germination (87.26%) was recorded which was higher among all records, while the lower seed germination (83.97) was recorded in seeds treated with liquid *A.vinelandii 1*. The germination percentage was increased due to inoculation of liquid bio-inoculants of *A.vinelandii* or may be due to ability to suppress the growth of antagonistic microbs present in soil or may be due to release of plant growth promoting substances around seed in rhizosphere.

Similar observations were also noted by Jadhav and Patil [11] reported that *Azotobacter* as biofertilizer performed better functions than inorganic fertilizers in comparison with seed germination of paddy plant. Sharma et al. [45] found that seed germination was increased due to inoculation of PSB in *Cicer arientinum* L. The bioinoculants *Azotobacter* and PSB found the significant increase in seed germination by Pathak et al. [44]. Similar findings by Mahato et al. [43] with *Azotobacter* biofertilizer. Simultaneously, Speed of Emergence (SE) and Seedling Vigor Index (SVI) were calculated after 6 weeks. More SE and SVI was observed by *A.vinelandii 3* while lowest SE and SVI was recorded by *A.vinelandii 1*.

## CONCLUSION

All the above-mentioned parameters of mung bean were influenced by *Azotobacter vinelandii*. Finding results from the above parameters are presented here and concluded significant increase in germination. Thus, growth and yield of mung bean were influenced by *Azotobacter vinelandii* which suit nutrient environmental conditions. Application of liquid *Azotobacter vinelandii* increase growth parameters and yield of lettuce crop significantly over uninoculated control. The significantly increased total *Azotobacter vinelandii* count in plant rhizosphere which further increased the available nitrogen and phosphorus content of soil. The uptake of nitrogen by *Vigna radiate* (Mung Bean) crop was also increased due to inoculation with liquid *Azotobacter vinelandii* bioinoculants. In addition to this, it stimulates rhizospheric microbes, defends the plants from harmful plant pathogens, increases nutrient uptake by the plants and improves the biological nitrogen fixation. The abundance of these bacteria in soil is related to many factors, mostly soil pH and fertility. Considering all these parameters, it could be concluded that liquid *Azotobacter vinelandii* improved soil biochemical properties which may ultimately influence the growth of *Vigna radiate* (Mung Beans) crop.

## ACKNOWLEDGMENT

The authors heartily acknowledge Dr. Jaiprakash S. Dargad, Principle, Dayanand Science College, Latur for his constant support and for providing Microbiology Research laboratory to work. The authors extends their gratitude towards Dayanand Education Society for the encouragement and providing financial support to carry out work. Also wants to extend thanks to Dr. S. S. Digrase, Senior Scientist of Krishi Vigyan Kendra, Latur for his assistance with authors. The authors also wish to acknowledge all respected teachers like Mrs. S. D. Patil, Dr. C. S. Swami including non-teaching staff for their constant support.

## REFERENCES

[1] Lawn R. J., Russell J. S., Mungbean: a summer grain legume for summer rainfall cropping areas of Australia. *Australian Journal of Agricultural Science*. 1978, 44: 28-41.

[2] Robertson M. J., Carberry P. S., Lucy M., Evaluation of a new cropping option using a participatory approach with on-farm monitoring and simulation: a case study of spring-sown mungbeans. *Australian Journal of Agricultural Research*. 2000, *51*:1-12.

[3] Cui S. Y., Yu D. Y., Estimates of relative contribution of biomass, harvest index, and yield components to soybean yield improvements in China. *Plant Breeding*. 2005, 124: 473-476.

[4] Duncan W. G., McCloud D. E., McGraw R. L., Boote K. J., Physiological aspects of peanut yield improvement. *Crop Science*. 1978, 18, 1015–1020.

[5] Gifford R. M., Thorne J. H., Hitz W. D., Giaquinta R. T., Crop productivity and photoassimilate partitioning. *Science*. 1984, 225: 801-808.

[6] Lal C., Hariprasanna K., Rathnakumar A. L., Gor H. K., Chikani B. M., Gene action for surrogate traits of water-use efficiency and harvest index in peanut (Arachi hypogaea). *Annals of Applied Biology*. 2006, 146: 165-172.

[7] Rao D. L. N., (2004). Edited paper on "Biofertilizer Research Progress work 2004-2007." All India Network Project on Biofrt., Indian Inst. of Soil Sci., ICAR, Bhopal.

[8] Chin M. W., (2010). The best organic liquid biofertilizer on earth replacing chemical fertilizers.Ezine Article.com http://ezinearticles.com/. The best organic liquid biofertilizer on earth replacing chemical fertilizer.

[9] Mishustin E. N., Naumova A. N., Bacterial fertilizers, their effectiveness and mode of action. *Microbiologiya.* 1962, 31: 543-555.

[10] Shende S. T., Rudraksha G. B., Apte R., Raut R. S., (1986). *Azotobacter* inoculation: Nitrogen economy and response of sorghum (CSH-1) cereal nitrogen fixation. In: *Proc. of working group meetings, held during 9-12 oct., 1984 at ICRISAT Center,* Patancheru, A.P., India.

[11] Jadhav S. W., Patil P. L., Effect of *Azotobacter* on growth and yield of paddy. *Indian J. Microbiol.* 1985, 21 (3): 212-213.

[12] Setubal J. C., dos Santos P., Goldman B. S. et al., Genome sequence of Azotobacter vinelandii, an obligate aerobe specialized to support diverse anaerobic metabolic processes. *J Bacteriol.* 2009, 191: 4534-4545.

[13] Ortiz-Marquez J. C. F., Nascimento M. D., Dublan Mde L., Curatti L., Association with an ammonium-excreting bacterium allows diazotrophic culture of oil-rich eukaryotic microalgae. *Appl Environ Microbiol.* 2012, 78: 2345-2352.

[14] Gimmestad M., Steigedal M., Ertesvaog H., Moreno S., Christensen B. E., Esp_ın G., Valla S., Identification and characterization of an Azotobacter vinelandii type I secretion system responsible for export of the AlgE-Type mannuronan C-5-epimerases. *J Bacteriol.* 2006, 188: 5551-5560.

[15] Kizilkaya R., Nitrogen fixation capacity of *Azotobacter*spp. Strains isolated from soils in different ecosystems and relationship between them and the microbiological properties of soils. *J Environ Biol.* 2009, 30(1):73-82.

[16] Vojinoviv Z., Microbiological properties of main types soil in Serbia for nitrogen cycling. *Journal for Scientific Agricultural Research.* 1961, 43:3- 25.

[17] Sariv Z., Biogenic levels of the horizons of calcerous chernozem in Vojvodina. *Contemporary Agriculture.* 1969, 17:819-825.

[18] Sariv Z., Biogenity of limeless chernozem in Vojvodina. *Annals of Scientific Work at the Institute for Research in Agriculture Novi Sad.* 1969, 7:145-151.

[19] Malek A. Y., Hosny I., Shawky B. T., Nitrogen–fixing capacity of *Azotobacter* as affected by the type and depth of substrate. *Zentralbl Bakteriol Naturwiss.* 1979, 134(5):390-397.

[20] Kalaigandhi V., Kannapiran E., Harimuraleedharan, et al., Azotobacter population in Rhizosphere and Non–Rhizosphere sediments of Tondi Coast. *International Journal of Biological Technology.* 2010, 1(1):63-65.

[21] Sariv Z., Ragoviv B., The influence of the maize on the dynamic of *Azotobacter* in the soil. *Soil Plant.* 1963, 13:273-277.

[22] Sariv Z., Ragoviv B., The effect of some plants on the dynamics of *Azotobacter* in the soil. *Annals of Scientific Work at the Faculty of Agriculture Novi Sad.* 1963, 7:1-11.

[23] Dong H., Li W., Eneji AE, et al., Nitrogen rate and plant density effects on yield and late–season leaf senescence of cotton raised on a saline field. *Field Crops Research*. 2012, 126:137-144.

[24] Monib M., Hosny I., Fayez M., Effect of *Azotobacter* inoculation on plant growth and soil nitrogen. *Zentralbl Bakteriol Naturwiss*. 1979, 134(2):140- 148.

[25] Rajaee S., Alikhani H. A., Raiesi F., Effect of Plant Growth Promoting Potentials of *Azotobacter chroococcum* Native Strains on Growth, Yield and Uptake of Nutrients in Wheat [2007–10]. *Agris records*. 2007, 11(41):285-297.

[26] Kizilog I. U., Bilen F. T., Ataplu N., Effect of Inoculation Eight *Azotobacter chroococcum* and Nitrogen Fertilizer on Plant Growth of Corn (*Zea mays*) Carbohydrate and Protein Contents. *Ziraat Fakultesi Dergisi Ataturk Universitasi*. 2001, 32:215-221.

[27] Meshram S. U., Shende S. T., Response of Maize to *Azotobacter chroococcum*. *Plant and Soil*. 1982, 69:265-273.

[28] Monib M., Malek A. E. Y., Hosny I, et al., Seed inoculation with Azotobacter chroococcum in sand cultures and its effect on nitrogen balance. *Zentralbl Bakteriol Naturwiss*. 1979, 134(3):243-248.

[29] Kizilkaya R., Yield response and nitrogen concentrations of spring wheat (*Triticum aestivum*) inoculated with *Azotobacter chroococcum* strains. *Ecological Engineering*. 2008, 33(2):150-156.

[30] Brakel J., Hilger F., Etude qualitative et quantitative de la synthese de substances de nature auxinique par *Azotobacter chroococcum in vitro*. *Bull Inst Agron Stns Rech Gembloux*. 1965, 33:469-487.

[31] Hennequin J. R., Blachere H., Recherches sur la synthese de phytohormones et de composes phenoliques par *Azotobacter* et des bacteries de la rhizosphere. *Ann Inst Pasteur*. 1966, 3:89-102.

[32] Azcorn R., Barea J. M., Synthesis of auxins, gibberellins and cytokinins by Azotobacter vinelandi and Azotobacter beijerinckii related to effects produced on tomato plants. *Plant Soil*. 1975, 43:609-619.

[33] Eklund E., Secondary effects of some *Pseudomonads* in the rhizosphere of peat grown cucumber plant. In: Pharis RP, et al. editors. *Hormonal Regulation of Development*. 1970, 3:613.

[34] Puertas A., Gonzales L. M., Aislamiento de cepasnativas de *Azotobacter chroococcum* en la provincial Granmayevaluacion de suactividadestimuladora en plantulas de tomate. *Cell Mol Life Sci.* 1999, 20:5-7.

[35] Althaf H. S., Srinivas P., Evaluation of plant growth promoting traits by *Pseudomonas* and *Azotobacter* Isolated From Rhizotic soils of two selected agro forestry tree species of godavari belt region, India. *Asian J Exp Biol Sci*. 2013, 4(3):431-436.

[36] Ben J. Tratado de Botánica Edic, Omega SA, Barcelona, España; 1964, 747.

[37] Brown M. E., Burlingham S. K., Production of plant growth substances by *Azotobacter chroococcum. J Gen Microbial.* 1968, 53:135-144.

[38] Mrkovac N., Milic V., Use of *Azotobacter chroococcum*as potentially useful in agricultural application. *Annals of Microbiology*. 2001, 51:145-158.

[39] Kamil P., Yami K. D., Singh A., Plant Growth Promotional Effect of *Azotobacter chroococcum*, *Piriformospora indica* and Vermicompost on Rice Plant. *Nepal Journal of Science and Technology*. 2008, 9:85-90.

[40] Anantha N. T., Earanna N., Suresh C. K., Influence of *Azotobacter chroococcum*strains on growth and biomass of *Adathoda vasica*Nees. *Karnataka J Agric Sci*. 2007, 20(3):613-615.

[41] Ravikumara S., Kathiresanb K., Thadedus M, et al., Nitrogen–fixing *Azotobacters* from mangrove habitat and their utility as marine biofertilizers. *Journal of Experimental Marine Biology and Ecology*. 2004, 312(1):5-17.

[42] Yasari E., Patwardhan A. M., Effect of (*Azotobacter* and *Azosprillium*) Inoculants and Chemical Fertilizers on Growth and Productivity of Canola (*Brassica napus* L.). *Asian Journal of Plant Sciences*. 2007, 6(1):77-82.

[43] Mahato P., Badoni A., Chauhan J. S., Effect of *Azotobacter* and nitrogen on seed germination and early seedling growth intomato. *Researcher.* 2009, 1(4):62-66.

[44] Pathak D. V., Singh S., Saini R. S., Impact of bio-inoculantas on seed germination and plant growth of Gauva (*Psidium guajava*) *J. of Hort. and forest.* 2013, 5(10):183-185.

[45] Sharma K., Dak G., Agrawal A., Bhatnagar M., Sharma R., Effect of phosphate solubilizing bacteria on the germination of *Cicer arietinum* seeds and seedling growth. *J. Herb. Med. and Toxicol.* 2007, 1:61-63.

In: Innovations and Challenges in Modern … ISBN: 978-1-53618-181-4
Editor: A. K. Haghi 

*Chapter 6*

# MICROWAVE IN PHARMACEUTICAL PROCESSES

***Hitesh V. Shahare[1,*] and Gokul S. Talele[2]***
[1]SNJB's Shriman Sureshdada Jain College of Pharmacy, Chandwad, MS, India
[2]NGSPM College of Pharmacy, Anjaneri, Nasik, MS, India

## ABSTRACT

Microwave radiations, an electromagnetic radiations, which are widely used a source of heating in organic synthesis. Microwave assisted organic synthesis (MAOS) is emerged as a new era in organic synthesis which makes the chemistry to go green for sustainable environment. The basic principle involved MAOS is Dipolar polarization and Conduction. This technique gives new opportunities to the synthetic researcher in the form of new reactions which are not feasible by conventional method, and it gives a flexible platform for carrying a various types of reactions. This technique gives a simple, clean, rapid and efficient way to researcher for synthesis of large number of molecules. This chapter focuses on principle,

* Corresponding Author's Email: hvshahare@gmail.com.

mechanism, various reactions by microwave, advantages, disadvantages and applications.

**Keywords:** green chemistry, green technology, microwave synthesis, microwave radiation, sustainable chemistry

## 1. History

Microwave is an accidental discovery. In 1946, the real potential of microwaves came to lime light when Percy le Baron Spencer, an American Engineer working with radar equipments while testing a new type of vacuum tube- the magnetron, at Raytheon noticed that some candy in his pocket, melted when he accidentally leant against an open wave guide (a rectangular hollow tube of metal used to conduct microwaves). Spencer began conducting more experiments with uncooked kernels and raw eggs. He watched popcorn bounce around the room and raw egg placed near the magnetron, explode from the pressure built up inside. The potential of Spencer's discovery was soon realized at Raytheon who began to produce the first microwave ovens called the "Radaranges". The Radaranges were large and expensive but it was not until 1967 when Amana, a subsidiary of Raytheon produced the first household microwave oven.

In the middle of 1980s, microwave ovens were used only for cooking and defrosting frozen food. However, in 1986, due to the independent work of Richard Gedye and co-workers at Laurentain University, Canada, microwaves successfully made the transition from kitchen to the chemical laboratory as they could do much more than mere cooking [1, 2].

### 1.1. Microwaves – Out of Kitchen into the Chemical Laboratory

The researchers reported increase in the rates of a number of organic reactions using a commercially available microwave oven. These reports formed the basis of an ever increasing range of research publications over the next twenty years. Later considerations for the safety and the need to achieve controllable, reproducible and focused homogenous heating led to the development of specially designed microwave equipments. The new microwave ovens were fitted with temperature and pressure detection devices making it possible to monitor a reaction while it is being irradiated under the oven [3].

## 2. INTRODUCTION

Microwave synthesis is the major breakthrough in the synthetic organic chemistry whereas the conventional heating is the inefficient and time-consuming. Microwave synthesis is the new lead which is being used as the source of heating in the organic synthetic reaction.

In 1855, Robert Bunsen invented the burner which acts as energy source for heating a reaction vessel; this was latter superseded by mantle, oil bath or hot plate, but the drawback of heating, though method remains the same.

Microwave assisted organic synthesis (MAOS) has emerged as a new "lead" in organic synthesis. The technique offers simple, clean, fast, efficient, and economic for the synthesis of a large number of organic molecules. Microwave can transfer energy directly to the reactive species, so called molecular heating; they can promote transformations that are currently not possible using conventional heat [4].

**Table 1. Conventional method & microwave method**

| **Conventional method of organic synthesis:** | **The Microwave system provides:** |
|---|---|
| Need longer heating time, | Green chemistry future |
| Tedious apparatus setup | Significant potential for a reduction of the by product |
| Result in higher cost of process | Lowering of the energy costs |
| Excessive use of solvents/ reagents lead to environmental pollution | Reduction in waste production |

# 3. Theory and Principle

A microwave is a form of electromagnetic energy that falls at the lower frequency end of the electromagnetic spectrum. It is in 300 to 300,000 megahertz frequency range. In this range, molecular rotations are affected and not the molecular structure. This places it between infrared radiation, which has shorter wavelength in the1-25 cm range for radar, whereas remaining section are devoted to telecommunication.

This microwave energy consist,

- An electric field and
- A magnetic field

Only electric field transfer energy to a heat substance, magnetic field interactions do not occurs in chemical synthesis.

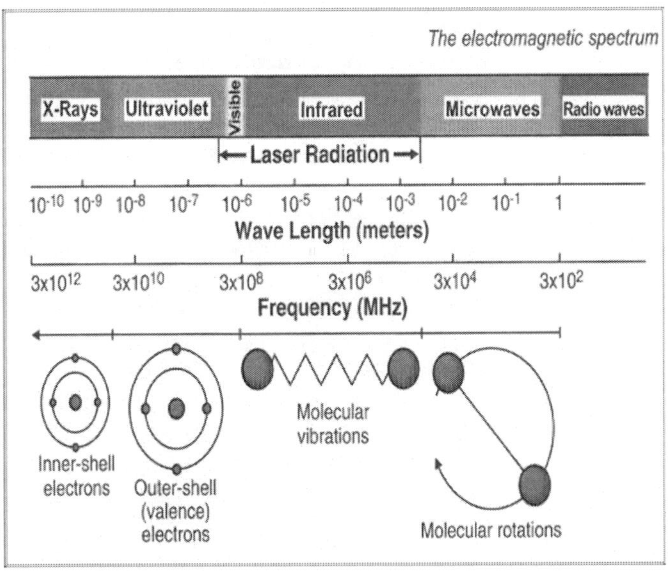


Figure 1. Electromagnetic spectrum and ranges.

Microwave moves with the speed of light and it have very less energy relative to the energy which is required to break the bond in the chemical molecule thus microwaves are such a source of energy which will not hamper the structure of the chemical molecule [5].

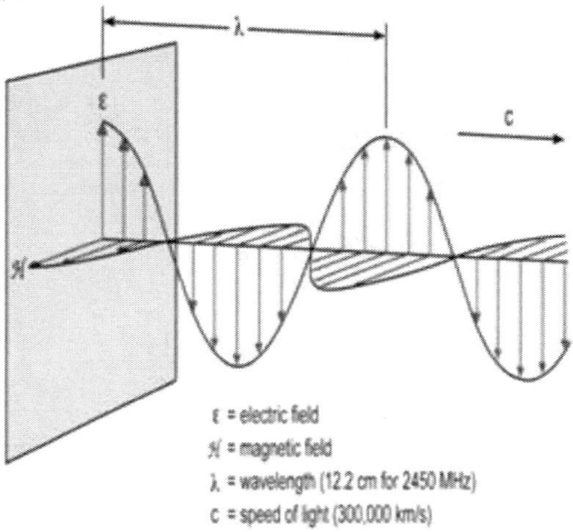


Figure 2. Microwave waves and energy.

## 3.1. Microwave Heating Verses Conventional Heating

Conventionally, chemical synthesis has been achieved through conductive heating with an external heat source, heat is driven into substance, passing first through the walls of the vessel in order to reach the solvent and reactants. This is a slow and in-efficient method for transferring energy into the system.

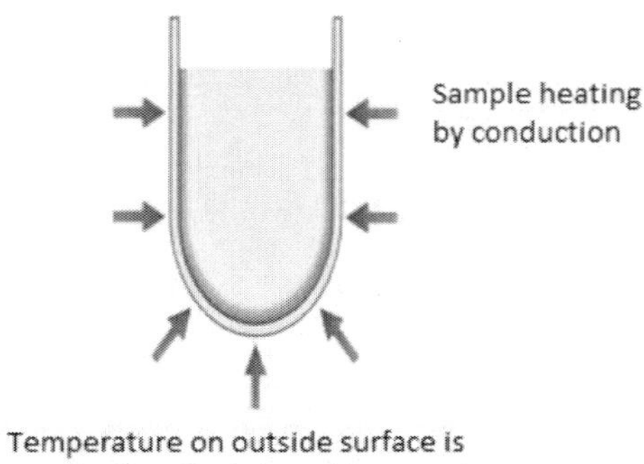

*Microwave heating* involves the direct heating of solution in vessel. This is a rapid and efficient method for transferring energy into the system.

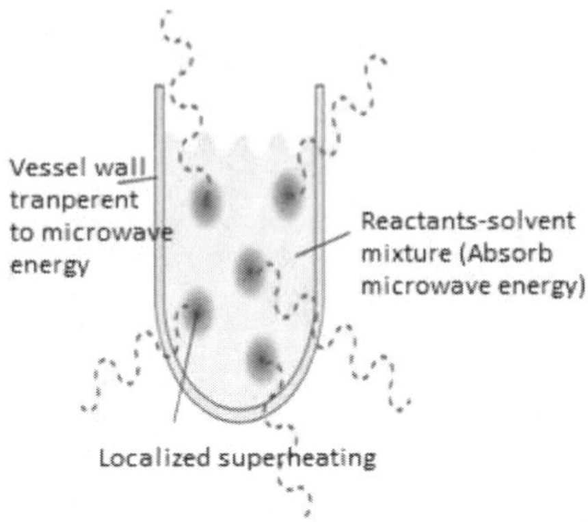


Sample heating by microwave

All the materials are not susceptible to microwave heating as response of various materials to microwave radiation is diverse. Microwave absorbing materials are most important for microwave chemistry, based on their response to microwaves, materials can be broadly classified as follows [6]:

1. Materials those are transparent to microwaves, e.g., Sulphur
2. Materials that reflect microwaves, e.g., Copper
3. Materials that absorb microwaves, e.g., Water

## 3.2. Principle

The basic principle behind the heating in microwave oven is due to the interaction of charged particle of the reaction material with electromagnetic wavelength of particular frequency. The phenomena of producing heat by electromagnetic irradiation are ether by collision or by conduction, some time by both.

All the wave energy changes its polarity from positive to negative with each cycle of the wave. This cause rapid orientation and reorientation of molecule, which cause heating by collision. If the charge particles of material are free to travel through the material (e.g., Electron in a sample of carbon), a current will induce which will travel in phase with the field. If charge particle are bound within regions of the material, the electric field component will cause them to move until opposing force balancing the electric force [7].

Two basic principal mechanisms involve in the heating of material;

1. Dipolar polarization
2. Conduction mechanism

These are the two fundamental mechanisms for transferring energy from microwaves to the substance being heated.

### *3.2.1. Dipole Interaction*

Polar ends of a molecule tend to align themselves and oscillate in step with the oscillating electrical field of the microwaves. Collisions and friction between the moving molecules result in heating.

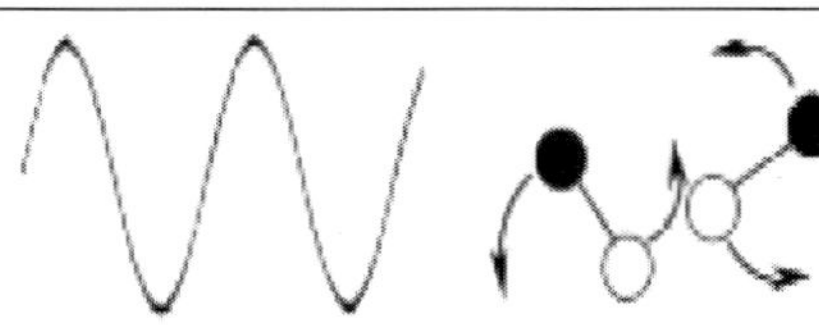

Dipolar molecules try to align with oscillating electrical field of microwaves

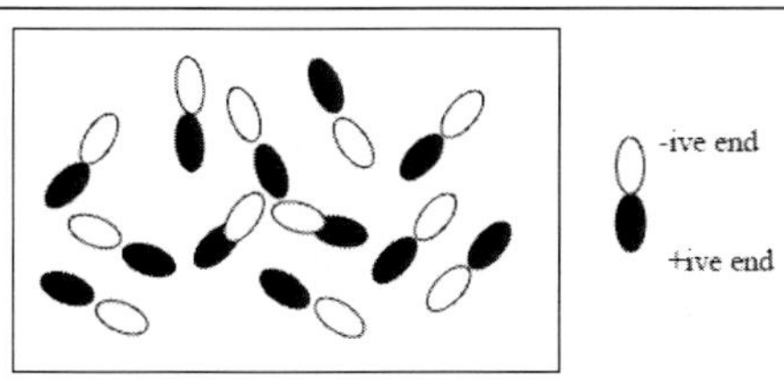


Dipoles in an absence of an electrical field

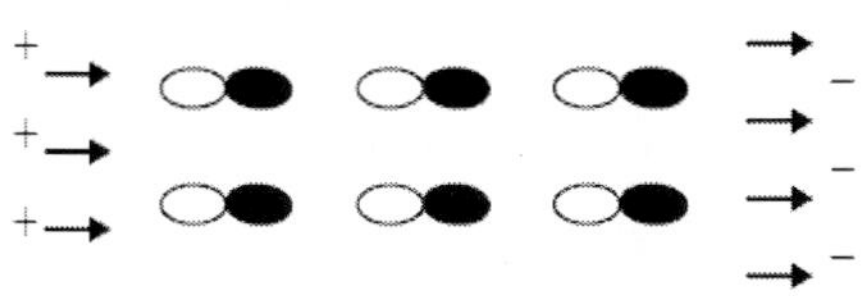

Dipoles in an presence of an electrical field

Figure 3. Dipole interaction in microwave.

*3.2.2. Ionic Conduction*

It results if there are free ions or ionic species present in the substance being heated. The electric field generates ionic motion as the molecules try to orient themselves to the rapidly changing field. This causes the instantaneous super heating [8, 9].

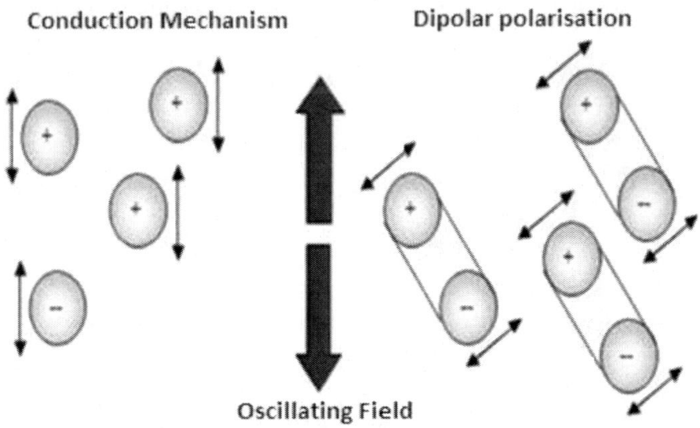


Figure 4. Ionic conduction in microwave.

## 4. Instrumentation

The apparatus for microwave assisted synthesis include [10]:

- Single-mode microwave ovens, and
- Multi-mode microwave ovens

### 4.1. Single-Mode Microwave Apparatus

The differentiating feature of a single-mode apparatus is its ability to create a standing wave pattern. This interface generates an array of nodes where microwave energy intensity is zero, and an array of antinodes where the magnitude of microwave energy is at its highest. One of the limitations of single-mode apparatus is that only one vessel can be irradiated at a time. However, the apparatus is user friendly.

An advantage of single-mode apparatus is their high rate of heating. This is because the sample is always placed at the antinodes of the field, where the intensity of microwave radiation is the highest. These apparatus can process volumes ranging from 0.2 to about 50 ml under sealed-vessel conditions, and volumes around 150 ml under open-vessel conditions. Single-mode microwave ovens are currently used for small-scale drug discovery, automation and combinatorial chemical applications.

## 4.2. Multi-Mode Microwave Apparatus

An essential feature of a multi-mode apparatus is the deliberate avoidance of generating a standing wave pattern inside it. The goal is to generate as much chaos as possible inside the apparatus. The greater the chaos, the higher is the dispersion of radiation, which increases the area that can cause effective heating inside the apparatus. As a result, a multi-mode microwave heating apparatus can accommodate a number of samples simultaneously for heating, unlike single-mode apparatus where only one sample can be irradiated at a time. A multimode heating apparatus is used for bulk heating and carrying out chemical analysis processes such as ashing, extraction, etc. In large multi-mode apparatus, several litres of reaction mixture can be processed in both open and closed-vessel conditions. A major limitation of multi-mode apparatus is that, heating samples cannot be controlled efficiently because of lack of temperature uniformity.

### *4.2.1. Components of Microwave Synthesizer*

Generally microwave oven consists of following parts:

1. A high power source
2. A waveguide feed
3. The oven cavity
4. Reaction vessel

5. Door interlocks
6. Exhaust fan
7. Time control
8. Mode stirrer

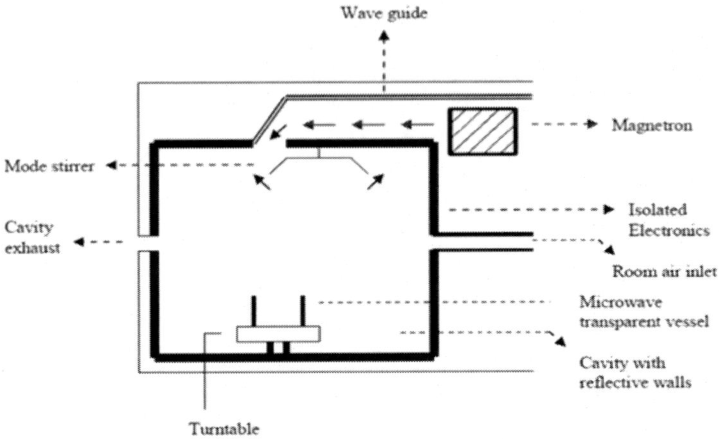


Figure 5. Microwave reactor.

1. *High power source:* A magnetron is a thermo ionic diode having anode and directly heated cathode which generate microwaves.
2. *Waveguide feed:* It is a rectangular channel which causes transmission of microwaves from magnetron to microwave cavity. It is have reflective walls made of sheet metal. These walls prevent leakage of radiations by increasing the efficiency of the oven.
3. *The oven cavity:* The area of oven cavity is designed in such a way so that it receives large amount of energy in the form of electric energy.
4. *Reaction vessel:* The reaction vessel for microwave induced organic reactions is a tall beaker, loosely covered which is having the capacity greater than volume of the reaction mixture. Teflon and polystyrene vessels can be used as these are transparent to microwaves. Metallic container gets heated soon due to preferential absorption and reflection of rays, hence those are avoided.

5. *Door interlocks:* These are safety devices in the door of the oven to prevent the door from being opened during microwave irradiation.
6. *Exhaust fan:* This isolates and ventilates the oven to prevent acid fumes from attacking the electronics of the unit.
7. *Time control*: This allows the time to be set for which the sample is to be irradiated.
8. *Mode stirrer:* A reflective fan shaped paddle to ensure that the microwaves are evenly distributed throughout the cavity.

## 4.3. Working of Microwave Oven

- In a microwave oven, microwaves are generated by a magnetron. A magnetron is a thermo-ionic diode having an anode and a directly heated cathode. As the cathode is heated, electrons are released and are attracted towards the anode. The anode is made up of an even number of small cavities, each of which acts as a tuned circuit. The anode is, therefore, a series of circuits, which are tuned to oscillate at a specific frequency or at its overtones.
- A very strong magnetic field is induced axially through the anode assembly and has the effect of bending the path of electrons as they travel from the cathode to the anode. As the deflected electrons pass through the cavity gaps, they induce a small charge into the tuned circuit, resulting in the oscillation of the cavity. Alternate cavities are linked by two small wire straps, which ensure the correct phase relationship. This process of oscillation continues until the oscillation has achieved sufficiently high amplitude. It is then taken off by the anode via an antenna.
- The variable power available in domestic ovens is produced by switching the magnetron on and off according to the duty cycle. Microwave dielectric heating is effective when the matrix has a sufficiently large dielectric loss tangent (i.e., contains molecules possessing a dipole moment) [11].

### 4.4. Conditions to Perform Microwave Synthesis[12, 13]

*Solvent* - Different solvents interact differently with microwave, because of their diverse polar and ionic properties.

*Volume* - Don't exceed a fall below the vials specific volumes

- Too low volumes: gives incorrect temperature measurement
- High volumes: does not leave sufficient heat space for pressure build up

*Concentration* - The concentration depends on the type of chemistry that is performed, the higher concentration gives faster reaction.

*Phase* - All different phases can be used, i.e., solution phase, solid phase, solid supported reagent, solvent free and scavenger resins

*Stirring* -Always add a magnetic stirring bar to the microwave vial.

*Inert atmosphere* - Inert atmosphere is not initially employed, if needed flush the vial with an inert gas before capping.

*Time* - most reactions require 2-15 min of irradiation.

*Temperature* - maintain the temperature between $60^0$C to $250^0$C

*Pressure* - the reaction can be safely performed at pressure of upto 20 bar.

## 5. Microwave Assisted Organic Reactions [MAOR]

MAOR are broadly classified into two categories [14, 15]:

- Microwave assisted reactions using solvent
- Microwave assisted reactions using solvent-free conditions

### 5.1. Microwave Assisted Reactions Using Solvent

The use of a solvent is not always mandatory for the transport of heat. The reactants are dissolved in the solvent which act as the energy transfer medium. The polarity plays major role as more in the polar reaction mixture. The reaction mixture will have better ability to couple up with the microwave energy. These are greener reactions because there is less or no use of catalyst, readily recyclable solvent and yield obtained is greater than conventional method. Various factors such as dielectric constant, dipole moment, tangential delta, dielectric loss and dielectric correlation will affect the polarity of the solvents and in turn affect the absorbing power of the solvent. Dielectric constant is also known as relative permittivity of the solvent, ability to store its electric charges.

The solvents used in Microwave Synthesis are:

- ✓ Water
- ✓ Formic acid
- ✓ DMSO
- ✓ Methanol
- ✓ Acetone
- ✓ Ethyl acetate
- ✓ Chloroform
- ✓ Toluene

### *5.1.1. Reactions in Microwave*

#### **5.1.1.1. Hydrolysis**

For example, hydrolysis of benzyl chloride with water in microwave for 3 min gives benzyl alcohol.

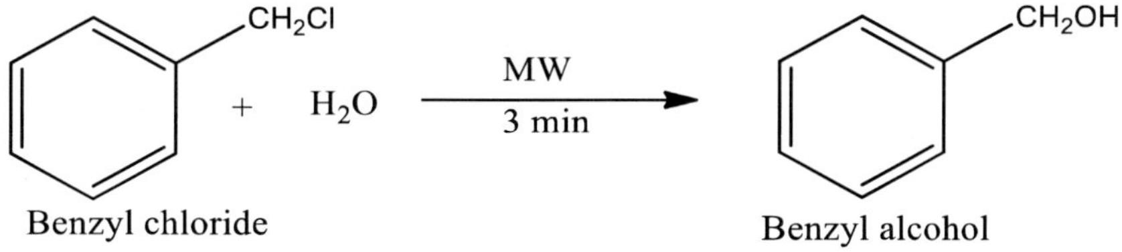

### 5.1.1.2. Oxidation

For example, oxidation of toluene with KMnO4 in microwave for 5 min gives oxidized product benzoic acid.

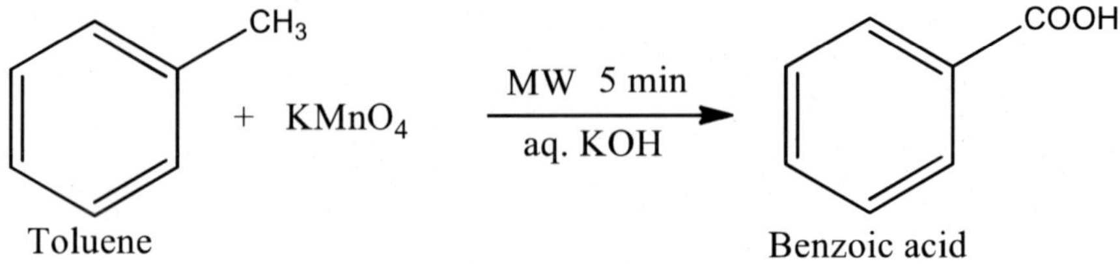


### 5.1.1.3. Cycloadditions

For example, cycloadducts are prepared by reaction between substituted amide and azide in toluene.

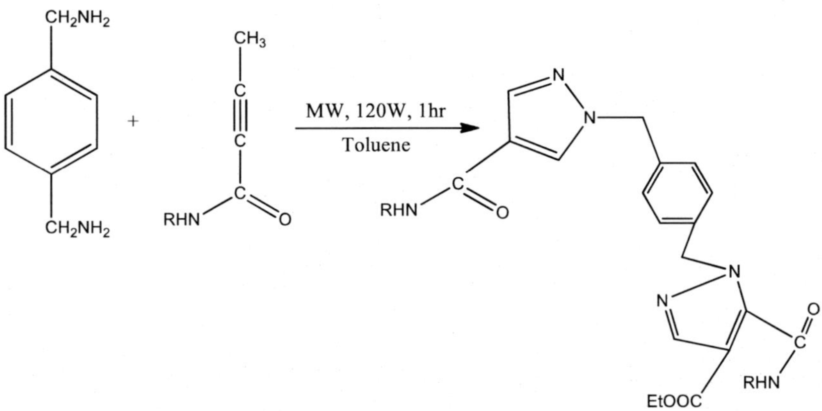


### 5.1.1.4. N-Alkylation

For example, N-alkylations of secondary amines and isocyanates in dichloromethane

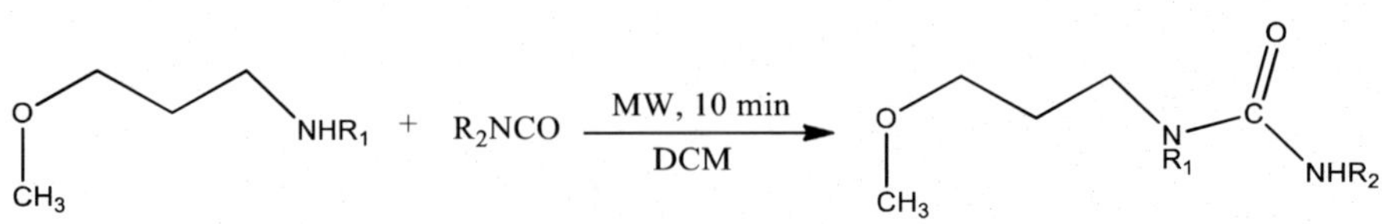

### 5.1.1.5. Esterification

For example, benzoic acid & n-propanol on heating in microwave for 6 min in presence of sulphuric acid gives propyl benzoate

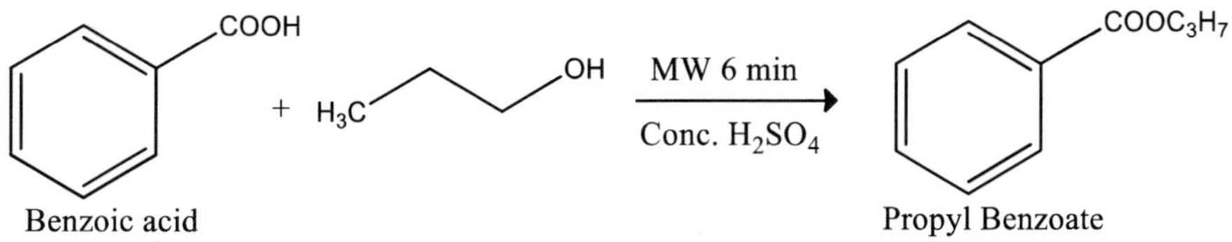


### 5.1.1.6. Decarboxylation

For example, decraboxylation by refluxing in quinoline in presence of copper chromate gives 6-methylindole in short time

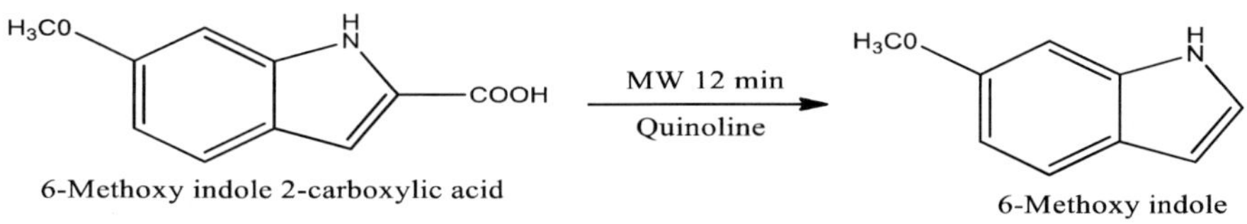


## 5.2. Microwave Assisted Reactions Using Solvent- Free Conditions

The reactions performed under solvent-free conditions present an alternative in the microwave chemistry and constitute an environmentally benign technique, which avoids the generation of toxic residues, like organic solvents and mineral acids, and thus allows the attainment of high yields of products at reduced environmental costs. This emerging environmentally benign technique belongs to the upcoming area of green chemistry.

Due to the environmental concerns, there has currently been an increasing demand for efficient synthetic processes and solvent-free reactions. Some old and new methodologies are being used to diminish and prevent pollution caused by chemical activities. Furthermore, microwave-assisted solvent-free organic synthesis has been developed as an environmentally friendly process as it combines the selectivity associated with most reactions carried out under microwaves with solvent and waste-free procedures in which organic solvents are avoided throughout all stages [16].

Solvent free organic synthesis are environmentally friendly process, these are of three types:

1. Reactions using neat reactants
2. Reactions using solid-liquid phase transfer catalysis (PTC)
3. Reactions using solid mineral supports

## 6. Advantages and Disadvantages [14]

### 6.1. Advantages

- Microwave assisted synthesis reduces the time of reaction substantially. Microwave enhancement may take several forms like reaction rate get accelerated, yields get improved than the conventional counterparts and virtually no decomposition takes place during the drying of samples.
- Microwaves form an essential aspect of green chemistry because of the solvent-free technique. Reactants can be adsorbed on solid supports like clay, montmorillonite, silica gel, alumina, etc and then exposed to microwaves. This eco-friendly procedure minimized the use of solvents leading to cleaner reactions and improved yields in addition to being safer. Ability to control the desired chemo, regio or stereoselectivity is possible using microwave assisted synthesis.
- Microwave heating can be used with less operator intervention, improved safety and greater control over the reaction conditions as well as minimum sample contamination and loss.
- Use of continuous flow microwave systems allows the samples to be digested or extracted in an online system for direct analysis.

- The advantages of microwaves are applicable to different disciplines of chemical research like drying of samples, melting of solid samples and a variety of organic and inorganic synthetic reactions.

### 6.2. Disadvantages

- Reaction requiring the use of dry nitrogen atmosphere, fuming, lachrymatory substances or substances which may corrode the interior of the oven cannot be conducted inside a microwave oven.
- There is a possibility that the higher temperatures/superheating of the solvent in sealed vessels may encourage the decomposition of the desired products or may lead to the formation of thermodynamically stable product in preference to the kinetically favored product.
- Metals are reflective to microwaves and the radiation tends to bounce off them like the light from a mirror. Due to this, metal particles or metals have to be avoided inside the microwave oven because there is always a possibility of an electric spark in the oven.
- No closed vessels should be used except the ones specially designed for withstanding high pressures like teflon tubes.
- The domestic microwave ovens, the power levels which significantly change from unit to unit.

## 7. Limitations of Microwave Assisted Synthesis

The yield obtained by using microwave apparatus available in the market is limited to a few grams. Although there have been developments in the recent past, relating to the scalability of microwave equipment, there is still a gap that needs to be spanned to make the technology scalable.

The use of microwaves as a source of heating has limited applicability for materials that absorb them. Improper use of microwave heating for rate enhancement of chemical reactions involving radioisotopes may result in uncontrolled radioactive decay.

Certain problems, with dangerous end results, have also been observed while conducting polar acid-based reactions, for example, microwave irradiation of are action involving concentrated sulphuric acid may damage the polymer vessel used for heating.

Conducting microwave reactions at high pressure conditions may also result in uncontrolled reactions and cause explosions.

Health hazards related to microwaves are caused by the penetration of microwaves. While microwaves operating at a low frequency range are only able to penetrate the human skin, higher frequency-range microwaves can reach body organs. Research has proven that on prolonged exposure microwaves may result in the complete degeneration of body tissues and cells.

## 8. Applications of Microwave Assisted Synthesis

Application of microwave irradiation in chemical synthesis involves its use in the acceleration of chemical synthesis. Microwave-enhanced synthesis results in faster reactions, higher yields, and increased product purity. In addition, due to the availability of high-capacity microwave apparatus, the yields of the experiments have now easily scaled up from milligrams to kilograms, without the need to alter reaction parameters. Microwave-assisted synthesis can be suitably applied to the drug discovery process.

### 8.1. Organic Synthesis [16]

Microwave-assisted organic synthesis has been the foremost and one of the most researched applications of microwaves in chemical reactions. The scientists have successfully conducted a large range of organic reactions. These include:

| | |
|---|---|
| • Diels-Alder reaction<br>• Heck reaction<br>• Suzuki reaction<br>• Mannich reaction<br>• Organometallic coupling reactions<br>• Cycloadditions reactions<br>• Heterocyclic chemistry<br>Dehydration | • Nucleophillic additions and substitution reactions<br>• Electrophilic substitutions<br>• Rearrangements<br>• Oxidations<br>• Reductions<br>• Condensations<br>• Hydrolysis<br>• Protection and deprotection |

Based on reaction conditions, organic synthesis reactions can be conducted in the following techniques;

1. *Microwave-assisted organic synthesis at atmospheric pressure*: Microwave-assisted organic synthesis can be most conveniently conducted at atmospheric pressure in reflux conditions, for example, oxidation of toluene to benzoic acid with $KMnO_4$ under normal conditions of refluxing takes 10-12 hours compared to reaction in microwave conditions, which takes only 5 minutes.
2. *Microwave-assisted organic synthesis at elevated pressure:* Microwaves can be used to directly heat the solvents in sealed microwave-transparent containers. The sealed container helps in increasing the pressure in the reactor, which facilitates the reaction that will take place at much higher temperatures. This results in a substantial increase in the reaction rate of microwave assisted organic synthesis.
3. *Microwave-assisted organic synthesis under solvent free conditions:* Microwave-assisted solvent-free organic synthesis has been developed as an environmentally friendly process as it combines the selectivity associated with most reactions carried out under microwaves with solvent and waste-free procedures in which organic solvents are avoided throughout all stages [17]

## 8.2. Inorganic Synthesis

A variety of materials such as carbides, nitrides, complex oxides, silicides, zeolites, apatite, etc. have been synthesized using microwaves.

- *Synthesis of organometallic and coordination compounds:* Microwave radiation has been successful in accelerating the reaction rate for the generation of organometallic and coordination compounds, which are produced by generating covalent bonds between organic compounds and metals.
- *Synthesis of intercalation compounds:* Applications of microwave chemistry for intercalation compounds have been tested. Intercalation compounds comprise organic or organometallic compounds that are incorporated between layers of oxides and sulphides. Conventional heating methods for the preparation of intercalation compounds, such as the intercalation of pyridine or its derivatives are slow and have limitations.
- *Synthesis of ceramic products:* In ceramic production industry, the removal of solvent or moisture is a critical step in the generation of ceramic products. Initially, the use of microwaves was limited to the effective removal of solvents from solid samples. It is estimated that for materials with a water content below 5%, microwave drying is efficient than conventional drying methods [18].

## 8.3. Synthesis of Nanotechnology Products

Amongst the several methods that exist for synthesizing of nanoparticles, the use of microwave assisted synthesis has shown promise. Synthesis of silver nanoparticles from silver nitrate employing starch as the reductant as well as stabilizing agent has been carried out under direct heating, controlled heating and microwave irradiation. The microwave irradiation was considered as better for reduction of silver ions to silver nanoparticles. It also afforded smaller particle sizes and particle size distribution. Compared to conventional methods, microwave assisted synthesis was faster and provided particles with an average particle size of 12 nm. Nanostructures with smaller sizes, narrower size distributions, and a higher degree of crystallization were obtained under microwave heating than those in conventional oil-bath heating. The gold nanoparticles have been prepared by microwave high-pressure procedure with alcohol as the reducing agent [19, 20].

## 8.4. Polymer Synthesis

Polymer chemistry, including ceramic processing, forms the single-largest application area of microwave chemistry. The use of polar reactants in polymerization reaction results in controlled synthesis and combination of this with direct heating of reactants makes microwave heating an economically viable option. The controlled solvent-free synthesis and modification in polymer materials can be rapidly and effectively done with the help of microwave heating using large scale reactors [21].

## 8.5. Synthesis of Radiopharmaceuticals

Microwave-assisted organic synthesis at an elevated pressure has been used in pharmaceutical industry for the synthesis of radiopharmaceuticals. During pre-clinical trials, these radiopharmaceuticals are used as tracers to generate a nuclear medical image. A multi-mode microwave oven was used in the first trial of this kind and it was observed that the rate of reaction increased substantially. This has resulted in the enhanced use of microwaves to produce radiopharmaceuticals.

## 8.6. Applications in Proteomics

Researchers began to apply microwave technology into proteomics so as to carry out enzymatic protein digestions. This non-classical proteomics method is identified to have enhanced digest efficiency and sequence coverage, and also accelerate degradation rates.

## 8.7. Automated Library Generation

Parallel set-ups are useful to carry out a microwave-assisted chemical reaction within a short time frame. However, there will be limited control over each reaction vessel regarding temperature/pressure. Each reaction vessel will be kept under same irradiation conditions in parallel set-ups. The reaction temperature is found to be associated with dielectric properties of solvent. So it is necessary to use same amount of identical solvent in each vessel so as to maintain temperature. The process of automated library generation is a viable alternative to parallel processing [18].

## 8.8. Solid Phase Organic Synthesis

Microwave-assisted synthesis is mostly embodied in synthesis procedure under solid-state condition. With the use of microwave irradiation, compared with traditional solution-phase chemistry, solid-phase synthesis in the presence of microwave irradiation has gained increasing attention from practitioners of combinatorial chemistry due to some advantages, like simplification of reaction work-up and product isolation, affinity and specificity of reaction rate-enhancements and the higher product yields [17].

## 8.9. In Analytical Chemistry

The various applications of microwave radiation in analytical chemistry are:

- Ashing

Ashing is used in various analytical laboratories to determine the ash content in a sample, for the purpose of process and quality control. Microwave heating is extensively used for ashing in the petroleum and fuels, plastics, pharmaceuticals and food industries.

- Digestion

Digestion is the process by which samples are broken down to their basic constituents for chemical analysis. Microwave digestion systems are used in analytical laboratories for sample decomposition and preparation. It involves the heating of microwave-absorbing reagents inside a pressurized, microwave-transparent container, in contrast to conventional open vessel digestion. Pressurization allows higher temperatures to be achieved in short period and it increases the speed of digestion.

- Extraction

Microwave extraction has proved to be more effective and efficient than its conventional counterpart, the Soxhlet extraction method. Extraction systems are used to conduct routine solvent extractions of soils, sediments, sludge, polymers and plastics, pulp and paper, biological tissues, textiles and food samples. Experiments have proved that microwaves, in comparison with the Soxhlet extraction, use a lesser volume of solvent and sample and perform extraction at a much faster rate.

- Moisture analysis

Application of microwave assisted moisture analysis has been extended to food and beverage, chemical, environmental, organic and pharmaceutical industries and has been found to be highly effective in reducing testing time. Microwave moisture analysis is specifically applied at product development stages such as process and quality control, testing of raw materials, intermediate and finished products [22].

## 9. Conventional Verses Microwave

**Table 1. Conventional method verses microwave method**

| | Conventional Method | | Microwave Method |
|---|---|---|---|
| 1 | Method is slow and inefficient method for transferring energy into reacting system | 1 | Method is rapid and transferring energy into the reacting vessel efficiently |
| 2 | Reaction mixture heating proceed via internal surface of vessel | 2 | Reaction mixture heating proceed directly into a mixture |
| 3 | Vessel should be in physical contact with the heating source | 3 | No need of vessel to be in physical contact with the heating source |
| 4 | Heating source is thermal or electrical | 4 | Heating source is electromagnetic waves |
| 5 | Heating mechanism is conduction | 5 | Heating mechanism is dielectric polarization and conduction |
| 6 | Heating rate is less | 6 | Heating rate is high |
| 7 | All reactants are heated equally | 7 | Specific component heated specifically |
| 8 | Highest temperature achieved is limited which is depends on boiling point of mixture | 8 | Temperature of mixture can be raised more than the boiling point. |

## Conclusion

The obvious features of microwave technology include carrying out solvent free reactions, reduction of time for a chemical/pharmaceutical reaction, instantaneous and uniform heating and possibility of parallel chemical reaction. This has proved as a bonanza for the researchers involved in drug discovery and development processes like high-speed combinatorial and medicinal chemistry. The combination of ionic liquids and microwave heating encourage scientists to initiate new unexplored areas of complex pharmaceutical systems. Microwave synthesis in macro-scale (synthesis of active pharmaceutical ingredient synthesis), micro-scale (integrated lab-on-achip" type approaches where synthesis and biological screening are integrated) and meso-scale flow units should be actively pursued.

## REFERENCES


[1] Loupy A. *Microwave in organic synthesis*. Weinheim: Wiley-VCH 2002; vol. 147, 180.

[2] Lancaster M. Green *Chemistry: An Introductory Text*, The Royal Society of Chemistry, London (2002).

[3] Zovinka E. P., Stock A. E. Microwave Instruments: Green Machines for Green Chemistry? *J. Chem. Educ.* 2010, *87*: 350–352,

[4] Joshi U. J., Gokhale K. M., Kanitkar A. P. Green chemistry: need for the hour. *Ind. J Edu. Res.* 2011; 168-174.

[5] Ley S. V., Baxendale I. R. New tools and concepts for modern organic synthesis, *Nat. Rev. Drug Discov.* 2002, 1(8): 573-586.

[6] Strauss C. R., Trainor R. W. Developments in Microwave-Assisted Organic Chemistry. *Aust. J. Chem.* 1995; 48(10): 1665-1692.

[7] Ravichandran S. and E. Karthikeyan. Microwave Synthesis - A Potential Tool for Green Chemistry. *International Journal of Chem Tech Research.* 2011; 3(1): 466-470.

[8] Mingos D. M. P, Baghurst D. R. Applications of microwave dielectric heating effect to synthetic problems in chemistry. *Chem. Soc. Rev.* 1991; 20: 1-47.

[9] Gabriel C., Gabriel S., Grant E. H., Mingos DMP. Dielectric parameters relevant to microwave dielectric heating. *Chem. Soc. Rev.* 1998; 27(3): 213-224.

[10] Kappe, C. O. *Practical Microwave Synthesis for Organic Chemists: Strategies, Instruments, and Protocols*; Wiley-VCH Verlag: Weinheim, Germany, 2009.

[11] Kappe C. O. Controlled microwave heating in modern organic synthesis. *Angew Chem Int Ed*. 2004; 43(46): 6250–84.

[12] Surati Madhvi, Jauhari Smita, Desai K. R. A brief review on Microwave assisted organic reactions. *Arch. Appl. Sci. Res*. 2012; 4 (1): 645-661.

[13] Gedye R., Smith F., Westaway K. et al. The use of microwave ovens for rapid organic synthesis. *Tetrahedron Lett.* 1986; 27: 279-82.

[14] Savita Belwal. Green revolution in chemistry by microwave assisted synthesis: A review. *Modern Chemistry*. 2013; 1(3): 22-25.

[15] Surati M. A., Jauhari S., Desai K. R. A brief review: Microwave assisted organic reaction. *Indian J. Pharm. Educ. Res*. 2012; 4(1): 645-661.

[16] Amit G. Nerkar, Dhanashri Pawale, Minal R. Ghante and Sanjay D. Sawant. Microwave assisted organic synthesis of some traditional and named reactions: A practical approach of green chemistry. *Int J Pharm Pharm Sci*. 2013; 5(3): 564-566.

[17] Chemat-Djenni Z., Hamada B., Chemat F. Atmospheric pressure microwave assisted heterogeneous catalytic reactions. *Molecules*. 2007; 12(7): 1399-1409.

[18] Rajak H, Mishra P. Microwave assisted combinatorial chemistry: The potential approach for acceleration of drug discovery, *J. Sci Ind Res*. 2004; 63(8): 641-654.

[19] Polshettiwar V., Nadagouda M. N., Varma R. S. Microwave-assisted chemistry: a rapid and sustainable route to synthesis of organics and nanomaterials. *Aust. J. Chem.* 2009; 62(1): 16-26.

[20] Ambrozic G., Orel Z. C., Zigon M. Microwave-assisted non-aqueous synthesis of ZnO nanoparticles, *Mater. Technol.* 2011; 45(3): 173-177.

[21] Hamid R. S., Mohsen S., Sudabeh R., Athar S. Glycerol as a biodegradable and reusable promoting medium for the catalyst free one-pot three component synthesis of 4H-pyrans. *Green Chem.* 2012; 14: 1696.

[22] Chen J. J., Deshpande S. V., Rapid synthesis of α-ketoamides using microwave irradiation–simultaneous cooling method. *Tetrahedron Lett.* 2003; 44(49): 8873-8876.

# ABOUT THE EDITOR

***Prof. A. K. Haghi, PhD***
Professor Emeritus of Engineering Sciences
University of Guilan, Iran
Email: AKHaghi@gmail.com; AKHaghi@Yahoo.com

A. K. Haghi, PhD, is the author and editor of 165 books, as well as 1000 published papers in various journals and conference proceedings. Dr. Haghi has received several grants, consulted for a number of major corporations, and is a frequent speaker to national and international audiences. Since 1983, he served as professor at several universities. He is former Editor-in-Chief of the *International Journal of Chemoinformatics and Chemical Engineering* and *Polymers Research Journal*; and is on the editorial boards of many international journals. He is also a member of the Canadian Research and Development Center of Sciences and Cultures (CRDCSC), Montreal, Quebec, Canada. He holds a BSc in urban and environmental engineering from the University of North Carolina (USA), an MSc in mechanical engineering from North Carolina A&T State University (USA), a DEA in applied mechanics, acoustics and materials from the Université de Technologie de Compiègne (France), and a PhD in engineering sciences from Université de Franche-Comté (France).

# INDEX

## A



## B

## C



## D

## E



## F



## G



## H



## I

## R



## S



## T



## U



## V



## W



## X



## Z